Oil, Wheat, & Wobblies

OIL, WHEAT & WOBBLIES

THE INDUSTRIAL WORKERS OF THE WORLD IN OKLAHOMA, 1905–1930

Nigel Anthony Sellars

UNIVERSITY OF OKLAHOMA PRESS
Norman

Sellars, Nigel Anthony
   Oil, wheat & wobblies : the Industrial Workers of the World in Oklahoma. 1905–1930 / Nigel Anthony Sellars.
       p.   cm.
   Includes bibliographical references and index.
   ISBN 978-0-8061-4327-9 (paper)
   1. Industrial workers of the World—History.   2. Trade unions—Oklahoma—History.   3. Petroleum workers—Oklahoma—History.   4. Agricultural laborers—Oklahoma—History.   I. Title.
   HD8055.I4S45   1998
   331.88'6—dc   97-30112
                                                                                       CIP

The paper in this book meets the guidelines
for permanence and durability of the Committee
on Production Guidelines for Book Longevity
of the Council on Library Resources, Inc. ∞

Copyright © 1998 by the University
of Oklahoma Press, Norman, Publishing
Division of the University.
Paperback edition published 2012.
Manufactured in the U.S.A.

All rights reserved. No part of this publication may be reproduced, stored in a retrieval system, or transmitted, in any form or by any means, electronic, mechanical, photocopying, recording, or otherwise—except as permitted under Section 107 or 108 of the United States Copyright Act—without the prior written permission of the University of Oklahoma Press.

*This book is for Nancy,
who kept me going*

The smallest effort is not lost,
Each wavelet on the ocean tossed
Aids in the ebb-tide or the flow;
Each rain-drop makes some floweret blow,
Each struggle lessens human woe.
      Covington Hall, *Rebellion*, May 1915

Contents

Acknowledgments xi
Introduction: Of Wobblies and Sooners 3
1 **Impossibilists, Reformers, and Labor Fakirs:** The IWW in Oklahoma, 1906–1914 15
2 **Harvesting the Harvesters:** The Agricultural Workers Organization in Oklahoma 35
3 **Organizing "Oily Willy":** The IWW in the Midcontinent Oil Fields 57
4 **With Folded Arms? Or with Squirrel Guns?:** The IWW and the Green Corn Rebellion 77
5 **War and Repression** 93
6 **Trials and Tribulations:** The IWW and the Red Scare in Oklahoma 119
7 **A Brief Renaissance:** Harvest Stiffs and Rednecks in the Postwar Period 143
8 **Decline and Fall, 1922–1930** 163
Epilogue: An Ambiguous Legacy 185
Notes 197
Bibliography 259
Index 283

Acknowledgments

THIS work grew out of my dissertation, which in turn was the product of both my long-standing interest in radicalism and socialism in Oklahoma and my family's background in trade unionism in England.

In the earliest stages of the dissertation I was greatly aided by Professor Norman Crockett, who encouraged me to read an essay based on my research at the Oklahoma Association of Professional Historians meeting in El Reno in 1988, and by Davis D. Joyce, of East Central University in Ada, who was inspired to publish a book on alternative views of Oklahoma history as a result of my paper. The essay subsequently appeared in that book, and I take responsibility for any errors in that work, errors that this book has sought to correct.

As the work proceeded, Professor Stephen H. Norwood, who chaired my doctoral committee, provided helpful criticism and allowed me to focus on the serious questions my research had revealed. I would also like to thank the rest of the committee— Norman Crockett, David W. Levy, H. Wayne Morgan, and David Gross—for their support, encouragement, and constructive criticism.

My research was aided by Raymond Boryczka, archivist at the Archives of Labor and Urban Affairs of the Walter P. Reuther Library at Wayne State University in Detroit, Michigan, and Edward Webber, curator of the Labadie Collection at the University of Michigan in Ann Arbor. I would also like to thank the staffs of the Western History Collections at the University of Oklahoma in Norman, the Oklahoma State

Archives in Oklahoma City, and the library and newspaper collections of the Oklahoma Historical Society in Oklahoma City. In addition, the interlibrary loan staff of the William Bennett Bizzell Memorial Library at the University of Oklahoma deserve my gratitude for the lengths to which they went to locate and obtain the numerous obscure and little-known newspapers, books, and journals that furthered my knowledge of the Industrial Workers of the World. Special thanks go to Brian Basore, a librarian at the Oklahoma Historical Society, who directed me toward some lesser-known material on labor and socialism held there; Steve Boyce, who shared material he had uncovered about the Oklahoma Wobblies; and Marsha Wiesiger and Sally Soelle, both of whom provided helpful criticism when I was just starting on this topic. Special thanks are due to Barbara Siegemund-Broka, whose careful editing of my manuscript caught grammatical errors large and small and helped make this a clearer, and therefore better, book.

I also want to thank Bob L. Blackburn, editor of *The Chronicles of Oklahoma*, his associate editor, Mary Ann Blochowiak, and the Oklahoma Historical Society for selecting this work in its early version as the best dissertation on Oklahoma history in 1994.

Financial support for this study came from the Henry J. Kaiser Family Fund of Menlo Park, California, which provided a grant that enabled me to do research at the Walter P. Reuther Library. The Graduate Student Senate and the Graduate College of the University of Oklahoma also provided conference grants that allowed me to read portions of this work at the annual conferences of the Southwest Social Sciences Association and the Western History Association.

Parts of this work have been either previously published or presented at professional meetings. Portions of chapters 3 and 5 were presented as "Wobblies in the Oil Fields: The Repression of the Industrial Workers of the World in Oklahoma" at the Oklahoma Association of Professional Historians annual meeting, held at El Reno Junior College, El Reno, Oklahoma, February 22, 1990, and as "The Suppression of the IWW in the

Sooner State," at the annual meeting of the Western Historical Association, Tulsa, Oklahoma, October 16, 1993. Chapter 4, "With Folded Arms? Or with Squirrel Guns? The IWW and the Green Corn Rebellion," was initially presented in a different form at the Southwest Social Science Association annual meeting, in Austin, Texas, March 21, 1992. "Wobblies in the Oil Fields: The Repression of the Industrial Workers of the World in Oklahoma" also appears in *An Oklahoma I Had Never Seen Before: Alternative Views of Oklahoma History*, edited by Davis D. Joyce and published by the University of Oklahoma Press in February 1994.

Last, but far from least, I want to thank my wife, Nancy Lee Sellars, for the help she has given me. With good humor she has endured my various joys and frustrations with this work and has provided much needed moral support at the times I felt more than overwhelmed.

NIGEL ANTHONY SELLARS

Norman, Oklahoma

Oil, Wheat, & Wobblies

Introduction
Of Wobblies and Sooners

JUST before midnight on November 9, 1917, a black-robed and hooded mob accosted sixteen prisoners. The prisoners, almost all members of the Industrial Workers of the World (IWW), were being escorted by nine Tulsa police officers from the city hall to the county jail. The so-called Knights of Liberty—including local businessmen, oil company executives, and police officials—took the prisoners at gunpoint and forced them into several cars, which then drove to a secluded ravine in the Osage Hills on Tulsa's west side. What followed was a hostile charivari, a vicious ritual in which the vigilantes beat, whipped, and tarred and feathered their captives, symbolically reducing them to the level of animals to be driven from the community of humans. "It was a party, a real American party," one newspaper called the incident, soon known as the Tulsa Outrage.[1]

With one act of savagery, the vigilantes used the patriotic fervor of the American entry into World War I to drive the IWW's Oil Workers Industrial Union from the Midcontinent oil fields of Oklahoma and southern Kansas. Motivated by the alleged bombing of an oil company official's Tulsa home and by inflammatory editorials in the local press, authorities also raided locals at Cushing and Drumright, completing the vigilantes' task. The Wobblies, as members of the IWW were nicknamed, faced further troubles when federal prosecutors, inspired by convictions stemming from Oklahoma's Green Corn Rebellion, began a series of conspiracy trials intended to destroy the radical labor union. Two of those trials—at Chicago and Wichita—included Wobblies with Oklahoma connections.[2]

But with the exception of the Tulsa Outrage and the Green Corn Rebellion, most historians of the IWW and Oklahoma radicalism have largely ignored the role of the Wobblies in Oklahoma. Historians of the IWW have focused on such dramatic incidents as the huge strikes in Lawrence, Massachusetts, and Paterson, New Jersey, various free-speech fights, and the numerous federal and state prosecutions of the union—an approach that has skewed IWW studies toward the East Coast and the Far West. Historians of Oklahoma radicalism have rarely strayed beyond the boundaries of the Socialist Party (SP) and have neglected other, equally powerful, influences. Yet the histories of the union and the state are more intertwined than one might imagine. For while Oklahoma nurtured the fortunes of American socialism, it was also where the Agricultural Workers Organization (AWO; later the Agricultural Workers Industrial Union, or AWIU), the financial and membership backbone of the IWW, began its recruiting campaigns and tested the tactics it would use throughout the harvest. It was in Oklahoma that the AWO, too long neglected by historians, began organizing the oil workers of the Midcontinent Field, an effort that became the IWW's major wartime recruiting campaign. And it was Oklahoma's dissident tenant farmers— though denied membership in the IWW—who drew more from the Wobblies' ideology and spirit than they did from SP campaign platforms when they organized the ill-conceived and ill-fated Green Corn Rebellion. Most important, both Oklahoma and the IWW were products of and responses to the culmination of a series of events dating back to the late nineteenth century, events shaped by the transformation of the American economy and the American West from a rural, agrarian society into a modern industrial one.[3]

This incorporation process introduced widespread and pervasive change into the national economy. Mechanization and massive economies of scale characterized the period, which saw the rise of national corporations and the displacement of small, regional companies. The new business enterprises, especially in the railroad and petroleum industries, soon came to dominate and shape the state and national governments to meet their own needs. Government no longer tried to encourage a "level playing field" for economic opportunity. Instead it focused on class-based legislation that favored the new corporate elite and undermined older agrarian and artisanal views of the American republic. The process also forever altered the traditional relationship between workers and employers as skilled workers saw their hard-earned skills and leverage in hiring and bargaining vanish. Mechanization began forcing all jobs down to the same unskilled or semiskilled level, and wages went with them—a process the IWW called "dequalification." Farmers, too, faced dequalification. Unable to compete with highly mechanized agribusiness and faced with selling their crops in a open market controlled by urban commodity brokers, they watched crop prices drop. With their lands heavily mortgaged, many found themselves forced into ruin, tenancy, or the unskilled labor force.[4]

In fact, the IWW and the state of Oklahoma represented two of the more radical forces of resistance to the emerging corporate order. Born of the struggles of western hard-rock miners and eastern mass production workers, the IWW fully recognized the direction this new economic and social transformation took and sought to control it for the benefit of the working class. But rather than looking backward to an idyllic—and largely nonexistent—past, the Wobblies turned to a bold vision of the future that could embrace traditional American values of equality and community as well as a respect for and fascination with the new technologies: the industrial union and syndicalism, ideas that in Oklahoma actually predated both populism and political socialism. Viewing political action as a retreat from class conflict at a time when they believed real change was possible, the Wobblies rejected that most sacred of political icons—the ballot box—as virtually useless. They fully recognized that the American system gave labor the right to vote but in it saw almost no room to maneuver for its own benefit. Only through direct confrontation with business could workers gain their rights, the IWW argued.[5]

Putting their beliefs into action, the Wobblies gained both fame and notoriety with their use of free-speech fights and mass strikes, such as the Lawrence textile strike of 1912. Their aggressive rhetoric, undiluted contempt for capitalism, and advocacy of sabotage also gave them an undeserved reputation for violence and lawlessness fed by lurid newspaper accounts of crop burnings, machine-wrecking, and dynamite plots. But their tactics and rhetoric also meant the Wobblies faced the wrath of increasingly powerful business interests, which utilized legislative, judicial, and extralegal means to suppress any opposition.

Similarly, Oklahoma came to be seen as a hotbed of radicalism, especially as residents learned the extent to which outside interests controlled the region's economy. Originally intended as a dumping ground for American Indians, Oklahoma by the 1880s became the farmers' last best hope, a refuge from the horrors of corporate capitalism and "wage slavery," a place where the republican ideals of equal opportunity, the work ethic, and individual economic success could still be realized. So powerful was this appeal that even the Knights of Labor called for opening Indian Territory to non-Indian settlement in order to keep Oklahoma out of the hands of cattle syndicates such as the Cherokee Strip Livestock Association, which the Knights saw as the advanced guard of corporate capitalism.[6]

Others shared this fear. When it entered the union in 1907, Oklahoma and its new constitution represented a direct challenge to the incorporators. Although the corporate business interests—represented in Oklahoma by the southern "business progressive" wing of the Democratic Party—controlled much of the state's economic life, they did not yet control its political structures. That remained in the hands of the resisters: old-style agrarians, populists of both radical and conservative stripes, socialists, and union labor. So strong was resistance to incorporation that Oklahoma's Socialist Party became one of the nation's largest, most electorally successful, and most influential in the years before the First World War. In Oklahoma, the resisters—like the IWW—shared a sophisticated understanding of the new economic order and an intrinsic distrust of corporations, and they wrote their fears into the state constitution. As a result, from statehood until well into the 1930s, the resisters and incorporators would do battle at all levels in Oklahoma—at the ballot box, in the courts and legislature, and on the shop floor. It was, as it turned out, a futile hope because Oklahoma, like much of the West, was already part of the new industrial order.[7]

As early as the Civil War, businessmen, bankers, and railroad promoters in Saint Louis and Kansas City, Missouri, and in Topeka and Wichita, Kansas, worked to develop industry within Indian Territory. By the 1880s cattle syndicates began attracting eastern and foreign capital to the region, while absentee business interests came to control the lead- and zinc-mining regions of northeastern Oklahoma and the lumber resources of the Choctaw Nation in the southeast. Whites who had married into the Five Tribes eagerly promoted the development of railroads in Indian Territory and the conversion of the region's natural resources into raw materials for profit. Agribusiness also emerged in Indian Territory as these same intermarried whites consolidated their land entitlements into plantations and hired tenants to grow cotton, while in Oklahoma Territory after white settlement smaller holdings—often lost though foreclosure—were merged into large wheat farms on par with the bonanza farms of the Northern Plains.[8]

The most important of all in the incorporating process were the railroads, which controlled the location of towns, owned the region's coal reserves, and even secretly funded efforts to open Indian Territory to white settlement. But while the railroads helped bring the businessman's frontier to the region, they also created the conditions that led to a wageworker's frontier. Prior to the railroad, labor in the West existed only in widely separated areas. The advent of the train meant more wageworkers went West, often drawn by the Horatio Algeresque promise of financial success and free land. Instead these workers learned that the great American dream was just that. As the the editor of the IWW's *Industrial Union Bulletin* noted in 1909, "But unlike the pioneer seeking a homestead and finding it, the modern wage-worker who 'goes West' finds no alternative except to hunt for a master."[9]

Many of these workers were first-generation citizens of the industrial society, much like the English workers on whose backs the Industrial Revolution was built. Many had made the journey from preindustrial society to the often incomprehensible world of industrial society and found themselves trapped between those two worlds. In addition, the extractive nature of the industries in which they worked, including agriculture, reduced many to the level of unskilled or semiskilled laborers just a step away from unemployment. This was in part because the railroads and mine owners preferred to strip an area for quick profits and relied on emerging technologies—which they controlled—that needed few, if any, skilled workers. What resulted was a pattern of exploitation and neglect in the West. In Oklahoma, this pattern continued well into the twentieth century, especially with the creation of oil-boom towns after 1910. With that pattern came the typical elements of the industrial frontier: company-owned mining towns with decrepit housing, company stores run on scrip, and a notable and deadly absence of safety regulations.[10]

But exploitation, physical and social isolation, and unsafe working conditions also created labor radicalism, especially of a syndicalist nature resembling the kind eventually espoused by the IWW. Like the Far West, these factors existed in the Oklahoma coal and metal mines, oil fields, and wheat farms. In the American West, workers grew alienated from the corporately controlled industrial world and developed deep-seated grievances against what they saw as a flagrant alliance of government officials and business interests that promoted profit at all cost and used force to silence dissent. These workers increasingly viewed economic growth as far from benign and argued that a society's worth depended as much on its moral values as on its capacity to produce goods. Precisely because they lacked any political power, the coal miners of Indian Territory, like their Rocky Mountain counterparts, turned to industrial unionism and a more syndicalist approach to express their discontent.[11]

While British and American miners from Pennsylvania brought a tradition of unionism with them, two Knights of Labor organizers from Illinois brought industrial unionism to Oklahoma in 1882. The two men, Dill Carrol and Frank Murphy, helped form district assemblies at Krebs, Lehigh, Coalgate, and other Oklahoma mining communities. Industrial unionism would also manifest itself in Oklahoma through Eugene Debs's American Railway Union, the United Mine Workers, and the American Labor Union, the latter a direct forerunner of the IWW.[12]

All these organizations espoused, in one form or another, an ideology with special resonance for workers in Indian and Oklahoma Territories. Indian Territory workers in particular were at the mercy of absentee employers: because they were not tribal members they were technically "guests" of the Indian nations and so lacked any political means to express their frustrations. Industrial unionism, with its emphasis on organizing all workers within an industry, helped create a sense of community among the workers, as opposed to craft unionism, which separated workers on the basis of skills.[13]

Industrial unionism also encouraged syndicalism, a doctrine that saw the trade union as the basis for a new, cooperative society. A hodgepodge of ideas, emotions, and militancy, American syndicalism was characterized by both a distrust of anything not emanating from the working class—especially middle-class reforms, which were thought to delay or diffuse revolutionary goals—and a contempt for established law and political action. As practiced by the otherwise reformist-oriented Knights and later by the IWW, syndicalism displayed hostility toward capitalism and toward bureaucracy, even in the union itself. It emphasized the workers' fullest control of their own working conditions and substituted on-the-job direct action, including sabotage, for the ballot box.[14]

In practical terms, syndicalism differed little from traditional trade unionism. It focused on improving working conditions, raising pay, and keeping job control in the hands of the workers rather than the newly emerging class of managers. As a revolutionary theory, though, syndicalism also sought to build a new society within the shell of the old. In Indian Territory and elsewhere, this included creating an alternative workers' subculture that functioned to overcome religious and ethnic differences among the workers and to compensate for a lack of stable social institutions. As a result, the union worked to establish cooperative stores, such as one planned for Krebs, and to set up a "court" system to deal with issues such as scabbing on fellow workers, accepting substandard wages, public intoxication, wife-beating, and desertion of a family.[15]

The system developed by the Knights essentially unraveled in Indian Territory with a failed national coal strike in 1894 and with the increasing impotence of the union's national leadership, which still sought to produce change through the ballot. Beginning in 1898, the United Mine Workers (UMW) picked up the mantle of industrial unionism in Oklahoma. While the Oklahoma UMW remained militant, with many of its members prominent Socialists, the union turned from direct action and toward the ballot box as Indian and Oklahoma Territories moved toward statehood. This change mostly resulted from the success of the miners' union and the Twin Territories' Federation of Labor in electing delegates to the constitutional convention and in writing portions of the document favorable to labor.[16]

But much of the state remained a wageworker's frontier. Conditions for workers remained poor, fostering labor militancy. Contemporary estimates by the Oklahoma State

Federation of Labor itself indicated that perhaps 99 percent of the state's workers were seasonal and unskilled, especially those in agriculture, the construction industries, and the nascent petroleum industry. They were unorganized, heavily exploited, migratory, and lacking in any political voice, Socialist or otherwise—precisely the sort of workers that the IWW sought to unionize. Even before Oklahoma became a state on November 16, 1907, some of those workers had already turned to the young labor organization to improve their lives.[17]

Although the IWW recruited only a few thousand dues-paying members from that group in Oklahoma, the Wobblies would still have tremendous influence. While more pragmatic unions needed large membership rolls and substantial dues to survive, the IWW saw itself equally as a propaganda organization for the coming revolution and as a union. Dues and memberships, while necessary, were less important than creating a sense of workers' solidarity and class consciousness. Recruiting among Oklahoma's harvest hands and construction and oil workers, IWW job delegates acted as evangelists for the struggle against the new corporate order, extending the union's message beyond its own membership. IWW appeals also influenced the state's tenant farmers, who were rapidly becoming a rural proletariat.[18]

But as the union extended its organizing efforts, it also encountered fierce resistance, especially in the oil-boom towns, the zinc- and lead-mining communities, and the blossoming agribusinesses of the Wheat Belt. The presence of IWW agitators raised fears in many communities that the discontent simmering below the surface of Oklahoma society might soon erupt. The Wobblies' radical rhetoric, the growth of the IWW's oil workers' union, and the Green Corn Rebellion of 1917 seemed to confirm these fears—and served to make the IWW a convenient whipping boy during the First World War. State officials, businessmen, and newspaper editors used the IWW as shorthand for allegations of anti-Americanism and treason, accusations that were also used to discredit and silence most other critics, especially the Socialists. But Wobbly harvest hands and roughnecks, more than any other radical group in Oklahoma or elsewhere, bore the brunt of hostile townsfolk and local authorities and incurred the wrath of oil company vigilantes, state antiradical laws, and the federal courts. Later they would face the more devastating effects of new technologies that totally eliminated the very jobs they performed.

The IWW eventually lost the battle in Oklahoma. As a consequence, one might be tempted simply to dismiss their efforts as a simplistic, backward-looking reaction to modernization. But clearly the tenant farmers, coal miners, harvest hands, and oil workers were the foundation of an emerging industrial working class in the region. Rather than try to restore older traditions, they attempted to cope with technological and economic changes of which they were both products and victims. The IWW was a logical outgrowth of those conditions, and its organizers spoke of resistance in terms with which workers could readily identify.

This book, then, seeks to remedy the general scholarly neglect of the IWW in Oklahoma, with particular reference to the AWO and to the ways in which the Oklahoma experience led the IWW to develop many of its distinctive recruiting tactics. In addition, I hope to demonstrate that the IWW's presence in the Oklahoma oil fields led directly to the massive state and federal efforts to prosecute and break the union, certainly a central event in the IWW's turbulent history.

This study also strives to place rank-and-file Wobblies and their work at the center of events. Although previous studies have focused on more colorful and highly publicized events such as mass strikes, free-speech fights, and high-profile trials, Wobblies were rarely made at such levels. They were made in the harvest fields and the pipeline camps, and their fates were determined not in federal conspiracy trials but by hostile townspeople, local laws against vagrancy and street-speaking, and state criminal syndicalism laws. The Oklahoma experience, in this respect, is one of the most illuminating.

INTRODUCTION

Finally, the fate of the IWW in Oklahoma may explain the decline of the Wobblies—and of labor in general—in the 1920s. Like the skilled workers of the American Federation of Labor, IWW members faced technological unemployment as new machinery displaced workers in the wheat and oil fields. The Oklahoma IWW fell victim both to its own failure to ally itself with groups such as the tenant farmers who shared its antipathy to corporate capitalism and to the various disputes over methods and ideology that effectively killed the union by 1924.

In one respect this work is a study of the IWW in one of its least-known and least-appreciated venues, but it also an attempt to demonstrate that radicalism in Oklahoma and the Southwest has its roots in a labor-based syndicalism. The history of the IWW in Oklahoma is part of a larger story, that of a nation undergoing profound economic, political, and technological change. It is also the tale of people who saw what they believed were American ideals being subverted by a new form of economic organization and who sought ways to control and humanize the industrial order.

1

Impossibilists, Reformers, and Labor Fakirs
The IWW in Oklahoma, 1906–1914

ON June 27, 1905, in Chicago's smoky, overcrowded Brand Hall, William D. Haywood, then vice president of the Western Federation of Miners (WFM), banged a piece of two-by-four on a podium and called to order "the Continental Congress of the working class." Over the next eleven days 203 delegates would create the Industrial Workers of the World. The new organization declared that no commonality of interest existed between employer and employee, embraced socialism, and extended membership to all wageworkers regardless of race, creed, color, or sex. It also rejected both the signing of contracts and affiliation with any political party, the latter a move that would create early schisms within the union.[1]

While convention attendees included representatives from several labor and left-wing political groups, the main impetus came from the WFM. Founded in 1893, the WFM grew from the bitter labor disputes in the silver- and copper-mining towns of the Rocky Mountain West. In the late 1890s and early 1900s, the union literally waged class war with mine owners and their political allies. Briefly affiliated with the American Federation of Labor (AFL), the WFM quickly realized that the AFL was interested neither in industrial unionism nor in organizing the West's unskilled workers. In response, the WFM in 1898 organized the Western Labor Union, which had little influence beyond areas dominated by the miners' union. It was reorganized as the American Labor Union (ALU) in June 1902. Possessing a strong commitment to political socialism, the ALU extended its influence beyond the Rocky Mountains and into

Oklahoma Territory, with locals in Guthrie, Oklahoma City, and Wanette by 1904. Consisting of small federal unions, or mixed locals, with members coming from several industries, the ALU seemed to be "coming to the front in Oklahoma Territory," declared the editor of the *American Labor Union Journal.*[2]

But the editor's optimism was misplaced as the ALU failed to grow and all of its locals broke up by 1905. Recognizing the ALU's weakness, the WFM in mid-1904 began working with other organizations and individuals—including Eugene Debs of the SP, "Mother" Mary Jones of the UMW, and the Socialist Labor Party's Socialist Labor and Trades Alliance—to create a new union based on an "Industrial Union Manifesto" issued in early 1905.[3]

But while Oklahomans took part in the ALU, their initial involvement with the IWW was more indirect. Some Oklahoma SP members and the more militant Indian Territory coal miners definitely expressed interest in forming a new labor organization. In fact, at the convention a UMW delegate from Pittsburg, Kansas, represented miners from Kansas, Indian Territory, Texas, and Arkansas who had earlier endorsed the WFM proposal. No delegates, however, came directly from Oklahoma, although two from Arkansas—Father Thomas J. Hagerty and Patrick ("Uncle Pat") O'Neil—were quite familiar to Indian Territory workers. Hagerty, a former Catholic priest and editor of the ALU's *Voice of Labor,* had organized railroad workers for the union just across the state line in Van Buren, Arkansas, and served as an Arkansas delegate to the 1904 SP national convention. O'Neil, a militant coal miner and a Socialist, was well-known and liked by Indian Territory UMW members. Both men played important roles in the early history of the IWW, with Hagerty helping create the union's structure and O'Neil organizing timber workers in Louisiana and eastern Texas and oil workers in Tampico, Mexico.[4]

Despite Oklahoma's initial lack of representation, the new IWW soon began to recruit in the region. By its second convention, in September 1906, the IWW had organized three locals in Oklahoma Territory and two in Indian Territory. The Shawnee Industrial Workers Mixed Union and the Oklahoma City Industrial Union received their charters in January 1906. Later, mixed locals at Perry in Oklahoma Territory and at Durant and Muskogee in Indian Territory also received charters. In addition, the IWW, operating through the WFM in Joplin, Missouri, began an unsuccessful recruiting campaign in the tristate mining region of Kansas, Missouri, and northeastern Oklahoma.[5]

A serious problem for the new IWW was that the union emerged after a massive upsurge in trade unionism, which had begun in 1898, ground to a halt in an economic downturn in 1903. The labor unrest that followed had allowed many industries—especially steel production and meatpacking—to break the unions in their midst. Oklahoma was not immune to such upheavals; the region's businesses developed as part of the national economy and not separate from it. Local businesses had either allied with eastern capital or depended on it for their survival, and they often adopted antilabor attitudes. Although the UMW had by 1903 gained union recognition and wage concessions, other workers faced stiff resistance. In Muskogee, the local builders' association, aided by a Citizens' Alliance, had locked out unionized construction workers in 1905, while an anti-union "open shop" movement appeared in Oklahoma City by 1906.[6]

Oklahoma workers also lacked the support of small-town business and professional elements, who in other states had shared hardships with local workers and usually retained more decent and humane social relations with their employees. Small towns in Oklahoma had virtually sprung up over night and lacked the organic development and social cohesion of those older towns. Entrepreneurs often controlled the communities, especially the oil-boom towns, and they desired quick profits and cheap labor. In addition, many Oklahoma middle-class reformers sought to impose morality and discipline on a chaotic society and viewed organized labor as a hindrance.[7]

In the face of such opposition, labor responded in several ways. The AFL's craft unions retreated from active recruiting and relied more on the union label on manufactured goods as a primary organizing tool. Some workers turned to the SP and to ballot box reform, but others rejected both the AFL's conservatism and the SP's electoral tactics and looked instead to the syndicalist approach of the "new unionism." While the IWW was committed to these goals, AFL unions such as the International Association of Machinists (IAM) and the United Brotherhood of Carpenters and Joiners (UBCJ) also moved toward this more radical position. Many IAM and UBCJ members, however, defected to the early IWW in part because they believed the AFL unions had failed to move toward syndicalism quickly enough. Among the first workers to encounter scientific management, machine technologies, and the anti-union open shop movement, the machinists responded with shop floor direct action, including sabotage, and quickly earned the nickname "war dog" of the AFL. The carpenters and joiners—especially under Peter McGuire, their longtime president and an avowed Marxist—had a history of militancy and opposition to the "business unionism" of Samuel Gompers, leader of the AFL.[8]

Not insignificantly, members of both unions controlled at least one early Oklahoma IWW local. While exact membership totals are not available, the Shawnee branch included five machinists and eight "wood butchers," and these men also composed the local's leadership. An Oklahoma Wobbly known only as "S." had particular praise for the men. Writing in the *Industrial Worker* he said, "They are generally the brightest men in the trades and quite independent; they are boomers, and when they loose a job they beat their way to another; they are missionaries for any principle they espouse."[9]

Carpenters and machinists probably composed part of the Oklahoma City and Muskogee locals as well, but available evidence suggests that the bulk of the Oklahoma City members were from the building trades. The city was undergoing a building boom at the time, and construction workers made up the largest group of unionized workers after the printing and railroad crafts. Again, exact membership figures are unavailable, but the local was large enough to spin off a separate Excavators Industrial Union local by September 1906. The composition of the other locals remains largely unknown but probably included railroad laborers as several of the towns were either railroad division points or, like Muskogee, possessed engine shops.[10]

Although the IWW probably tried to recruit African Americans and women, white males dominated the organization nationally and in Oklahoma. Outside of the UMW, few blacks worked in unionized industries, and many white workers harbored the racist and nativist views of the era. Despite their rhetoric, some Wobblies also shared those prejudices, which almost certainly accounts for the nearly complete lack of African-American and Native American members in Oklahoma. By the same token, at least some male Wobblies still accepted nineteenth-century, middle-class ideas about women and cast female workers in a domestic role. Many workers, however, infused those values with a working-class spirit and opposed capitalism because it forced women to work. Union members could then show their manhood and the strength of their union by their ability to protect their wives and children from wage labor. The existence of a women's auxiliary to the Muskogee local—a rare, if not unique, occurrence for the IWW—clearly reflects the traditional role of a supportive wife aiding her union husband.[11]

In contrast, women took leadership roles in the Oklahoma City local. Those who joined the local may have been seamstresses, dressmakers, or laundry workers, all common types of employment for working-class women in the city. But one member, Ethel E. Carpenter, was the wife of a physician and was more of a middle-class reformer. She and another woman, Della Weinstein, led the branch's defense committee, which raised funds for three WFM officers—Haywood, Charles Moyer, and George Pettibone—accused of murdering a former governor of Idaho. Carpenter—whose husband, J. D. Carpenter, lectured for the IWW—also contributed articles to the *Industrial Worker*. One piece was the first to urge the IWW to organize the harvest migratories who eventually became the financial and membership backbone of the union.[12]

Some male Wobblies, such as Oklahoman Frank Little, came to appreciate the fighting spirit of female workers, who saw themselves exploited both directly as industrial workers and indirectly as the wives and mothers of workers. In 1916, Little called on the IWW actively to recruit women and to develop both a separate bureau and agitators to deal specifically with their needs. But many Wobblies also believed women should leave existing moral and social standards unchallenged until capitalism ended and workers achieved economic freedom. Only then, they argued, would gender and family roles be equalized. Complicating matters, the migratories who came to dominate the union believed their very lack of family and job ties made them more radical. As a result, the IWW made only a few efforts to organize women. Outside of the Oklahoma City local, the only other evidence of Oklahoma women in the IWW came in 1917 when Mrs. Elmer F. Buse of Tulsa wrote Bill Haywood about a Denver, Colorado, domestic workers' union organized by Jane Street. Street responded to Mrs. Buse's letter, but nothing resulted from the inquiry.[13]

While the roles of African Americans and women in the IWW were generally issues of concern within the union, the IWW also had to deal with the region's AFL and the growing Oklahoma Socialist Party. Almost from the moment the IWW appeared in Oklahoma, the union encountered resistance from both groups. In general, the IWW's disputes with the SP and the AFL in Oklahoma resembled those the union had in other states. The Wobblies' emphasis on industrial unionism threatened the craft union structure of the AFL, while the Socialists—who favored industrial unionism—saw the IWW's rejection of the ballot box as undermining the party's legislative efforts. SP leaders such as Victor Berger and Oklahoman Oscar Ameringer declared as "impossibilist" any attempt to use revolution to build socialism. Both groups condemned the Wobblies' support of sabotage, which they said drove people away from unionism and political socialism.[14]

Yet the IWW's position on industrial unions and politics grew naturally from the working conditions of the West, conditions replicated in Oklahoma. As the IWW saw it, industrial unionism was not just revolutionary, it also proved more effective for "bread-and-butter" objectives such as higher wages and better working conditions. The union's refusal to scorn those immediate goals meant it differed from the AFL only in degree, not kind. Similarly, most western workers were migratory and could not meet residency requirements to vote, which the IWW argued rendered the ballot box useless.[15]

In fact, AFL-style craft unionism also opposed corporate control by creating labor organizations that drew on a definite working-class consciousness. Like the allegedly more radical Wobblies, the AFL leadership rejected the reform-oriented, ballot box approach and concerned itself with job conditions and wages. But while Samuel Gompers believed craft unions were the best means to achieve these goals, the IWW opposed the AFL's emphasis on craft exclusivity, arguing that it threatened working-class solidarity and was out of step with the new mass production industries that made skilled labor unimportant. In particular, IWW ideologists blamed the craft system and its apprenticeship program for restricting access to the trades. As a result, workers frozen out of the trades had only one alternative—to become strikebreakers. The Wobblies additionally viewed the existence of "union scabbing"—that is, the continued work of one craft union while another was on strike in the same factory—as a direct result of separating workers by craft. "Union Scabs," a scathingly satirical pamphlet on this topic written by Oscar Ameringer, was especially popular with the IWW. Although he personally opposed the IWW, Ameringer actively supported industrial unionism and declared that union scabbing was caused by craft organization.[16]

For their part, many Oklahoma AFL leaders accepted bread-and-butter unionism and identified with the values of a business class that was better organized than labor. The Oklahoma State Federation of Labor (OSFL) viewed the Wobblies as a competing or "dual" union that threatened the OSFL's legislative successes. That fear may have also derived from the union leaders' realization that only 15 to 18 percent of Oklahoma workers were unionized, and that fully one-third of those were members of the UMW, an industrial union with a history of militancy.[17]

Because of the strength of the miners, the question of industrial unionism was far from an easy one for the OSFL. Several important federation officials, especially J. Luther Langston and Victor Purdy, strongly advocated industrial unionism and would eventually bolt from the OSFL in the 1930s to lead the state's Congress of Industrial Organizations. Socialists within the federation also criticized craft unions. In addition, while critical of the Wobblies' ideas, some Oklahoma AFL members gave financial and moral support to IWW strikes, such as the Coalgate UMW local whose members petitioned Secretary of Labor William B. Wilson in 1916 urging the release of Wobblies and iron miners jailed "at the bidding of the Steel Trust" in Minnesota's Mesabi Range.[18]

That did not stop some Oklahoma leaders from attacking industrial unionism. The printing crafts were especially critical, mostly as a result of their jurisdictional disputes with the industrially organized International Typographical Union. The *Oklahoma Labor Unit*, a semi-official OSFL paper owned by the Oklahoma City Printing Trades Council, regularly ran editorials alleging that industrial unionism would produce ruin if forced prematurely on workers. The *Labor Unit*'s editors, C. C. Zeigler and Ollie S. Wilson, also never hesitated to attack the IWW, labeling it "the pirate of the industrial sea" and its members "false prophets" out to destroy the concept of right and wrong, pervert the meaning of justice, and create a despotic dictatorship. The editors on occasion drifted into a conspiratorial mode of thought, suggesting that the IWW was really a front for anti-union businessmen out to disrupt the AFL. The attacks grew especially shrill after an alleged attempt by Oklahoma City Socialists to commandeer the city's central trades council in 1912 in the wake of a failed 1911 streetcar strike. The printers accused a leading insurgent, H. C. Waller, of being a Wobbly, although no evidence exists that Waller belonged to the IWW.[19]

Opposition to the IWW also occurred at the local level. Immediately after its formation, the Shawnee branch ran into trouble with the city's AFL affiliates and the trades council. T. F. Delaney, a machinist elected the IWW local president, faced a series of charges from the Shawnee IAM local, but the charges were dropped when Delaney said he would use the occasion to plead the IWW cause. The trades council president, allegedly acting on orders from Samuel Gompers himself, also tried to oust carpenter E. E. Matteson from the assembly. But Matteson argued that the trades council had no power to eject him as his AFL union had elected him its representative. Later attempts to expel Wobblies from AFL locals such as the plumbers' union and the UMW were more successful, especially during World War I.[20]

The Oklahoma City IWW, however, encountered on-the-job resistance from the AFL. In one incident in the summer of 1906, an anonymous correspondent told the *Industrial Worker* that a sympathetic foreman had hired IWW members to rebuild part of a local meatpacking business damaged by a fire. The AFL local had allegedly refused the job because it was on a ten-hour-per-day basis. Once on the job, the IWW men, with the foreman's help, promptly agitated for the eight-hour day and began an effort to organize the packinghouse workers. As a result, the foreman was fired, and the Wobblies walked off the job. The construction company agreed to the eight-hour demand, but the AFL union took the contract for themselves. The IWW members were dismissed and attacked as scabs.[21]

The IWW faced similar opposition from the state's Socialist Party, although some party members sympathized with the union. Despite a reputation for radicalism, the Oklahoma SP more often adopted the approach of Victor Berger and other relatively conservative party leaders, a strategy based on success at the ballot box, ameliorative legislation, and "boring-from-within" in an effort to take over the AFL. Radical SP members generally derided the Berger program as "slowshulism," but even Oklahoma's left-wing party members showed little enthusiasm for the IWW's more revolutionary stances. This was in part because the Oklahoma Socialists had had some success in working with the AFL, and some important members of the OSFL—especially J. Luther Langston and Ira Finley—were either Socialists themselves or at least sympathetic to the SP goals.[22]

Still, some party members, particularly in the revolutionary or "Red" wing, did support both the Wobblies' form of industrial unionism and its tactics. Even before the birth of the IWW, A. B. Davis of Geary, who had helped form the Oklahoma Territory Socialist Party in 1899, served as a member of the ALU's propaganda brigade. Oklahoma Socialists also took part in IWW free-speech fights in Spokane, Washington, and San Diego, California, including Bruce Rogers of Ada, who served as the IWW's attorney in Spokane. Another party member, J. F. "Frank" Ryan, who had taken part in an SP free-speech fight in Tulsa in 1914, later helped organize an IWW oil workers local in Tulsa in 1917 and was one of the victims of the Tulsa Outrage.[23]

The Wobblies also had support from several state Socialist newspapers, including H. Grady Milner's *New Century* of Sulphur, Jasper Roberts's *Beckham County Advocate*, and the Oklahoma City *Social Democrat*, especially when it was edited by Stanley J. Clark. A major figure in the Oklahoma left wing, Clark had briefly held IWW credentials while organizing in Texas. Clark's strong positions apparently worried the federal government because he was later convicted in the mass Chicago IWW trial of 1918 even though he no longer belonged to the union. Like Clark, Jasper Roberts sympathized with the IWW.

In the wake of the SP's national debate on the use of sabotage, Roberts praised Bill Haywood, then general-secretary of the IWW, for refusing to accept crumbs from employers just to gain political advantage. He called Haywood America's leading revolutionist and criticized those who claimed the Wobblies were too radical. "Sometimes I think we are not radical enough," Roberts wrote.[24]

Another Oklahoma Socialist, E. J. Foote, became a significant figure in the early history of the IWW. Foote, a cook and a baker, had lengthy experience in the SP and knew Oklahoma well. In 1904, he had lived in Medford, where a Socialist Labor Party (SLP) local had formed in 1895. He was on good terms with many of the older Medford radicals and briefly edited the *Medford Challenge*, a Socialist newspaper that regularly ran articles by Father Hagerty and other left-wingers. He also contributed articles urging direct action to the first incarnation of the *Industrial Worker*. By 1906, he had moved to Wichita, where he successfully maneuvered the Wichita local of the AFL Bakery and Confectionery Workers Union into the IWW. Foote later went to Washington State, where he took part in the IWW's November 1909 Spokane free-speech fight along with fellow Oklahoman Frank Little.[25]

Yet despite such support, the Oklahoma SP as a rule opposed the IWW and sided with the conservative majority on two major issues: the Socialist Party's rejection in 1912 of sabotage and the 1913 recall of Bill Haywood from the National Executive Committee, which grew out of that stance. The issue of sabotage, also known as direct action, proved a critical one because it colored popular mythology about the Wobblies as violent and destructive. A relatively new word, "sabotage" quickly came to mean violence, usually against property, especially machines. This meaning developed from a widespread but erroneous belief that the word derived from striking French workers throwing their wooden shoes, called *sabots*, into machinery to destroy it, and because labor strife in the United States had a history of violence. Oklahomans, and not just workers, had engaged in just such destructive acts during the 1886 Great Southwestern Railroad Strike and the 1894 Pullman Strike, occasionally with fatal results. Given such incidents, it was easy to see how Wobbly talk of sabotage invoked images of destruction, especially among firms dependent on machines for their economic survival.[26]

But sabotage as the IWW meant it was far more complicated and far less violent. The Wobblies drew on the original meaning of the word, that of French workers who pushed their workplace demands by acting like the clumsy, unskilled, and wooden-shoed country folk who often acted as strikebreakers. The general idea was to pretend to be stupid or uncomprehending, thereby slowing down production to the point where it cost the employer profits. Similar action was well-known elsewhere: in England, it was called "rattening" or "skulking"; in Scotland, "ca' canny." In the United States, it was known as "soldiering"—the term used by Frederick W. Taylor, a proponent of scientific management who opposed such practices—or as "playing the Hoosier," again referring to strikebreakers from rural areas. The Wobblies also used another term, "to strike on the job," a clear and concise expression of the intent of sabotage.[27]

For the IWW, then, sabotage possessed a broad meaning, including passive resistance. It was any tactic that "hit the employer where it hurt: in the pocketbook," as one IWW strategist termed it. Tactics could include telling customers that the employer's products were poorly made, following a foreman's order to increase production by running a machine at high speed until it seized up, or by simply folding one's arms and refusing either to work or to leave until demands were met. All of these were common workplace strategies and were not limited to unionized workers, as early studies by John R. Commons, a labor researcher, and Stanley Mathewson, a management specialist, showed. The Wobblies, therefore, came under attack for openly advocating what most workers already did.[28]

But even though the IWW rejected violence and was generally scrupulous in using nonviolent tactics in strikes, the confrontational nature of its rhetoric worried the SP. The "Yellow," or right-wing, socialists saw sabotage as an anarchistic threat to the party's electoral successes. The left-wing "Reds" feared sabotage because employers could retaliate with greater physical force and because many workers held onto older, agricultural views that venerated private property. To endorse sabotage, they argued, was to court repression from employers and alienate the workers themselves. Oklahoma Socialists such as Oscar Ameringer and George E. Owens, editor of the party-owned *Oklahoma Pioneer*, joined in the condemnation.[29]

The issue came to a head in May 1912 when the party's national convention amended the SP constitution to expel any party member who advocated crime, violence, or sabotage. It was clearly aimed at the IWW. Despite opposition from some left wingers, the amendment passed by a 191-to-90 vote and was later overwhelmingly endorsed by the party membership. Supporting the resolution was the majority of the Oklahoma delegation. Only one Oklahoma delegate opposed the amendment: John G. Willis, a blacklisted member of Eugene Debs's old American Railway Union. In a May 25, 1912, *Oklahoma Pioneer* editorial, Owens labeled anarchist anyone who advocated sabotage and declared that they threatened the orderly, ballot-box progression to socialism. "The Socialist party is not a mad mob of desperate men moved by hatred to the commission of all forms of violence and destruction."[30]

The amendment became the basis for a recall of Haywood, who had been elected to the National Executive Committee in 1911. Party conservatives, who disliked Haywood, attacked him for supporting sabotage in a December 1, 1912, speech in New York City and claimed he had rejected political action, which was untrue. In fact, like many Wobblies, Haywood supported legislation on child labor and minimum wages. But the right-wing Socialists quickly moved to recall Haywood from the committee, and they succeeded in February 1913, when party members approved the recall, voting 22,495 to 10,944. Although left-wingers such as Stanley Clark and J. T. "Tad" Cumbie had captured the state party machinery, and several socialist newspapers—such as the *Social Democrat* and the *Grant County Socialist*—actively defended Haywood, most Oklahomans voted with the majority, with 539 members supporting the recall versus 294 backing the IWW leader.[31]

But the final result of the sabotage debate was the overall weakening of the American radical movement. Those left-wingers who deserted the SP over the expulsion effectively handed the party to the reformists, while those who deserted the IWW denied the Wobblies the intellectual ability, theoretical insights, and commonsense restraint on reckless action that the union needed. In addition, the sabotage amendment would come back to haunt the SP. Ironically, because the party became the first group to create a definition of permissible belief and action within the framework of discontent, they inspired state criminal syndicalism laws, including Oklahoma's, and provided the federal government with the rationale for prosecuting radicals during World War I.[32]

But what doomed the first IWW presence in Oklahoma was neither confrontations with the AFL nor with the SP. Instead, two internal union schisms and an economic downturn in 1907 did the job. The first split occurred at the union's second convention, in September 1906, when a faction led by William Trautmann, the union's secretary-treasurer, SLP leader Daniel DeLeon, and Vincent St. John, a militant WFM organizer, succeeded in ousting IWW president Charles O. Sherman. Sherman was accused of misappropriating seven thousand dollars in union funds and of trying to drive radicals out of the union. The two Oklahoma delegates—William O'Donnel, the Oklahoma City local's delegate, and E. J. Foote, who served on a committee to rewrite the IWW constitution—both sided with the insurgents. Sherman formed a short-lived rump IWW at Joliet, Illinois, and, in a fairly futile gesture, expelled those

locals—including Oklahoma City Industrial Union no. 239—
that failed to support him.³³

Local 239 did not fare well. In April 1907, its new leadership
had thanked Trautmann for sending a former coal miner
named Clinton Simonton to help with recruitment, but by
September 1907 E. J. Foote, then in Wichita, urged the IWW's
General Executive Board (GEB) to send a job delegate to
Oklahoma City and three other state towns to organize bakery,
hotel, and restaurant workers. "I promised this to the bakery
workers in Oklahoma City when I organized [there,] and I fear
that they will make very slow progress," he wrote Trautmann.³⁴

But the GEB was then dominated by DeLeon supporters led
by Michael Katz. This group hoped to move the IWW toward
more overtly political positions and to reduce it to simply the
labor union arm of the SLP. As a result, the GEB showed little
interest in Foote's proposal. They agreed to pay for any Wichita
local member who wished to go to Oklahoma City but rejected
Foote's request to send Lillian Forberg, who had organized coal
miners in Kansas. The GEB said Forberg's work was "not up to
standard," but the real reason may have been her support in
1906 for the ousted Sherman, even though she had remained
with Trautmann's faction. Whatever the case, nothing came of
Foote's efforts. Additionally, the nation was in the grip of the
Panic of 1907 and Oklahoma City's building boom came to an
end. By 1908, no IWW locals remained in Oklahoma.³⁵

That year produced the IWW's second serious schism, when
DeLeon and his supporters clashed with the more straight-
forward unionists over the need to subordinate political action
to economic action. A group of migratory workers from the
Pacific Northwest, labeled the "Bummery" by DeLeon, joined
with eastern opponents of DeLeon to force DeLeon and his
allies from the convention. An angered DeLeon declared the
"new" IWW anarchist and formed his own version, headquar-
tered at Detroit. Dominated by SLP members and largely
existing only on paper, the Detroit IWW—later renamed the

Workers International Industrial Union (WIIU)—led a handful of strikes before the SLP eventually reabsorbed it. The only time the WIIU appeared in Oklahoma was 1919, when an oil worker named "M. H." from Lenapah in northern Nowata County wrote the union's *Industrial Union News* urging it to recruit in the oil fields. The tiny organization, however, was clearly incapable of such a campaign.[36]

After the 1908 convention, a distinct change in the composition of the IWW arose. Membership increasingly shifted from urban mass production workers and toward "floaters," unskilled migratory laborers such as lumberjacks, harvest hands, and oil workers. This change was significant for Oklahoma because large numbers of its workers were migratories. In fact, the numbers of migratories increased so much after statehood that the Oklahoma State Department of Labor omitted statistics on wage earners in its 1910 report, saying it was too difficult to obtain information from the same individuals each year.[37]

One of the best-known IWW organizers of the migratories had strong Oklahoma ties: Frank Little, often considered the quintessential Wobbly. Born in Illinois to a Quaker physician father of English extraction and a southern mother who was one-eighth Cherokee, Little moved with his family in the 1890s to Oklahoma's Payne County, where his older brother Alonzo had staked a claim in the 1889 Land Run. Little briefly attended Oklahoma Agricultural and Mechanical College at Stillwater, but left school after his father's death in January 1899 and migrated to the copper mines of Colorado. He later joined the Western Federation of Miners, becoming part of Vincent St. John's pro-IWW faction.[38]

Tall, muscular, and blind in one eye, Little took a leading role in major IWW free-speech fights in Kansas City, Spokane, and Fresno, California. Like Father Hagerty before him, Little rejected the ballot box completely and considered himself at war with the existing economic order. In 1911, his uncompromising stance helped elect him to the union's General Executive Board, a post he held until his August 1917 murder by vigilantes in

Butte, Montana. Little made several organizing trips to Oklahoma and became a popular speaker. But between 1908 and Little's return to the state in November 1911, the IWW had managed to organize only a single, short-lived local at Wagoner, probably composed of migratories doing railroad construction work. An organizer from Kansas City named Tom Halcrow had also failed to revive the Oklahoma City local in early 1911.[39]

Little arrived in Oklahoma just after taking part in the Kansas City free-speech fight. Ill at the time, he stayed with a friend, a Mrs. Allie L. Cox, in Guthrie, but wrote to *Solidarity*, then the main IWW newspaper, to report on conditions. By January 1912, Little had recovered enough to resume organizing. He revived the Oklahoma City local and later helped organize an oil workers' local in Drumright. Little also gained popularity with the state's tenant farmers, many of whom already had experience with organizations tending toward syndicalism such as the Oklahoma Renters Union and the Land League of Oklahoma and Texas. Covington Hall, a Louisiana IWW member, noted in his memoirs that Little, protected by armed guards, often spoke at moonlit meetings of Socialist Party locals. Many of the farmers apparently agreed with Little's rejection of the ballot box as a means of change. "[T]hey don't see how the socialists will be able to carry on production; how they will be able to construct the new government," Little told *Solidarity*.[40]

While many tenants wanted to join the IWW, they were disappointed when Little told them only wage earners could belong to the union. Although denied membership, many farmers—undoubtedly influenced by Little's speeches—later joined a syndicalist organization of their own, the Working Class Union (WCU). Ironically, it was the WCU that brought the hand of repression down on the Wobblies, although the IWW had few connections to the tenant organization and had even rejected its requests for affiliation.[41]

But in the long run Little's efforts fell short. By December, the Oklahoma City local had disbanded, and from 1913 to 1914 no IWW locals existed in Oklahoma. A few individual Wobblies did report to IWW papers on working conditions in Oklahoma, but the union lacked an organized presence until an emerging oil boom and the state and federal government's botched attempt to control the wheat harvest job market gave the IWW renewed life.[42]

Although it existed only a short while, the early Oklahoma IWW still provides much information on both the original composition of the union and the high degree of economic and technological change in the nation. The presence of militant members of the International Association of Machinists and the United Brotherhood of Carpenters and Joiners in the state's locals indicates how urban and industrialized Oklahoma had become even while it remained a wageworker's frontier for the unskilled and mobile. It also suggests that these workers' experiences differed little from those of workers in the East. The lack of African-American and female members indicates that these white male workers also maintained long-standing racial and gender prejudices, despite the influence of women in the Oklahoma City local.

Similarly, the Oklahoma IWW, like its national and state counterparts, had to deal with opposition from the AFL and from the SP. While much of this conflict concerned dual unionism or use of the ballot box versus outright revolution as the road to change, the disputes in Oklahoma are interesting precisely because one would have expected the more radical positions of the Oklahoma State Federation of Labor and the state SP, indeed the state in general, to have been conducive to the IWW. But the OSFL fought the IWW precisely because the federation had achieved some legislative successes, which the IWW possibly threatened. In addition, with only a portion of the state unionized, the OSFL feared workers might find the IWW position appealing and defect, weakening the AFL in the state while increasing employers' attacks on the union movement. Similarly, the Socialists feared the IWW's aggressive rhetoric would threaten the party's electoral gains and erode its rank-and-file strength. While opposition was strongest among the more moderate, reformist wing, the otherwise sympathetic left wing also feared that the IWW's calls for sabotage would bring down the wrath of employers. Ironically, by supporting the antisabotage amendment and the recall of Haywood, the Oklahoma Socialists may have actually weakened their own political position and allowed the state's conservative, incorporating forces to gain the upper hand.

But the Wobblies' internecine squabbling proved an equally important factor in the demise of the first Oklahoma IWW locals. The splits alienated potential allies and eventually placed the union in the hands of a more militant, migratory faction. In Oklahoma, this probably pushed some workers back to the AFL. When coupled with the Panic of 1907 and the decline of the building boom in Oklahoma City and other towns, the urban labor movement's shift toward caution and craft exclusivity becomes readily understandable. As a result, the IWW failed to grow in the settled industries and increasingly found itself involved with workers in the extractive industries, such as petroleum and agriculture, where economies of scale were just taking hold. While the domination of the IWW by the "floaters" occurred because the union failed to hold the settled workers, eventually the leadership elected to embrace the migratory workforce, a policy encouraged early on by Oklahoma Wobblies such as Ethel Carpenter, E. J. Foote, and Frank Little. By 1914, the boom towns and wheat fields of Oklahoma would begin to serve as testing grounds for recruiting those men whom the IWW then believed were the vanguard of the revolution.

2

Harvesting the Harvesters
The Agricultural Workers Organization in Oklahoma

BY 1914 the IWW as a whole was a relatively weak organization outside the lumber industry in the Pacific Northwest. Even its brilliant orchestration of the 1912 textile strike in Lawrence, Massachusetts, had failed to produce a stable local among the mostly immigrant workers, and the union had followed up that success with a disastrous strike of silk workers in Paterson, New Jersey. In addition, the recall of Haywood by the SP had strained the union's relations with other radical organizations, although some left-wing Socialists briefly swung over to the IWW.

By 1913, the IWW had about 14,310 dues-paying members—less than one-hundredth the membership of the AFL and one-twentieth that of any industrial union in the AFL. Many union members had already begun to ask, "Why doesn't the IWW grow?" The answer came almost by accident when, in 1914, job recruitment policies in the Wheat Belt produced an unmanageable labor surplus for the wheat harvest. The overabundance of workers created widespread unrest and violence throughout the Midwest. By September 1914, the IWW leadership realized that the migratory harvesters, or "hoboes," represented a vast pool of potential members. In fact, union leaders began to see the migratory workers as "the franc tireurs of the class struggle."[1]

Seizing the initiative, the IWW in 1915 began one of its most successful recruitment campaigns and its most profitable branch union, the Agricultural Workers Organization no. 400. The campaign demonstrated that unskilled migratory labor could be effectively organized, and it turned the Oklahoma Wheat Belt town of Enid into a test area for such hallmark Wobbly tactics as roving job delegates; the short, quick strike; and the strike on the job. The recruiting drive also overcame tremendous obstacles based on middle-class fears of radicalism and unionism, the very nature and conditions of unskilled work, and the attitudes and behaviors of the migrants themselves. As a result, the IWW learned to emphasize more "bread-and-butter" issues such as wages and hours and to deemphasize the union's revolutionary rhetoric. While their efforts brought the Wobblies new strength and financial resources, the IWW grew too dependent on the harvest hands and paid a high price when newer technology arrived in the wheat fields.

Although men, and a few women, had tramped to find work since colonial times, the armies of migratory wheat harvest hands were the product of the partial rationalization of agriculture at the end of the nineteenth century. Before the Civil War a few urban artisans might have joined the harvest to earn money when their own work was slack. But the increasing use of mechanical reapers after the war and the growing presence of railroads helped create a business-oriented, economy-of-scale agriculture that required large numbers of seasonal workers.[2]

Wheat culture came late to Oklahoma, partly due to a lengthy drought in the late 1890s. Between 1913 and 1919 Oklahoma produced 66 million bushels of wheat annually, accounting for 90 percent of all grain cultivated in Oklahoma and 80 percent of the state's farm income. The area around Garfield County in north-central Oklahoma marked the start of the massive wheat farming region, which ran north to the Dakotas and southern Canada. Ten rail outlets served Enid, the county seat, which each May became the initial gathering point for men seeking to follow the harvest to its autumn end in Manitoba and Saskatchewan.[3]

Although Oklahoma had some giant agribusiness operations, known as "bonanza" farms, which were larger than 640 acres (such as the 101 Ranch near Ponca City), medium-sized farms of at least 320 acres produced the bulk of the harvest. While smaller operations might get by on help from family and neighbors, most Oklahoma wheat farms required outside labor to bring in the grain. A 1921 United States Department of Agriculture survey noted that only 19 percent of farms of 320 to 400 acres could operate without extra hands, and that number dropped to 10 percent for farms larger than 640 acres. As a result, migratories composed 64.2 percent of the state's harvest workforce.[4]

The need for workers depended also on the type of machinery used. Smaller wheat farms used "binders," which cut the unripened wheat and bound it into shocks. This required only two to four men—called "shockers"—to follow the machine, collect the bundles, and turn them upside down to ripen. The work proved slow and back-breaking because the binder could harvest at most ten to twelve acres a day.[5]

By comparison, machines called "headers" cut more than half the wheat in north-central Oklahoma. Unlike the binder, the header only cut the heads of ripened grain and did not bind them. Because the ripened grain could shatter and scatter its seeds, header operations required a large number of men working intensely over a short period of days to harvest the grain. Header crews included men who drove the horse-drawn reapers, men who drove wagons called "barges" that transported the harvested grain, and laborers who arranged the wheat in the wagons for transport and in "ricks" for storage. Headers could cut about thirty acres a day, producing about a week's worth of work on a given farm. After the harvest some migratories could also find additional work with a threshing crew, although most moved on to areas where the harvest was just starting.[6]

The work itself was hard, with conditions spartan at best. Even travel to the fields could be deadly. Oklahoma newspapers regularly ran stories about young men killed while trying to hop freights for the Wheat Belt. Brakemen often threw hoboes from trains, while "hi-jacks"—robbers who preyed on the migratories—probably murdered others.[7]

Once in the Wheat Belt, migrants learned that accommodations were few, and many had to stay in "jungle camps" outside the town. Ralph Chaplin, the IWW's best-known poet and editor, said the harvesters slept in haystacks, in empty boxcars, or in the open. The hoboes scrounged food, such as boiling wheat for gruel or to make flour for pancakes, or concocted a "hobo's delight" from young alfalfa and bacon rinds. "What made John Farmer mad was to find fresh vegetables piled near the 'jungle' fire, or a plump rooster simmering in the stew pot. Sometimes things like that were hard to explain away," Chaplin said.[8]

Finding a job often proved difficult. While farmers sometimes came to the jungles, both sides more commonly engaged in "curbstone bargaining," with the farmers in wagons and the "bindlestiffs" on the sidewalks. Supply and demand set wages, with pay somewhat less in Oklahoma than in Kansas; and supply usually exceeded demand. Most men worked as many as ten jobs in a row to make any money at all. Expenses often ate into earnings and reduced any hope of saving a "stake" to last through the jobless winter. One Oklahoma farmer who made the harvest in 1919 found only 69½ hours of work for the entire harvest. He earned $31.27 but incurred $20.00 in expenses.[9]

Working conditions were often miserable. The harvesters labored in the intense heat and humidity of summer, and many suffered sunstroke and heat exhaustion. Jack Miller, a harvest hand who joined the IWW in 1916, recalled seeing five men go down with heat exhaustion in a single day. He added that the work and heat were so taxing that fat men were few in the fields. Workers also inhaled grain dust and other particles, especially when working near a reaper or threshing machine. While a thunderstorm might offer some respite from the heat, the accompanying lightning posed a serious safety hazard.[10]

Danger also lurked in the machinery itself. Careless harvesters could lose a hand or foot to the reaper's razor sharp sickles or be killed outright, or a barge might tip over, crushing the driver. Grain dust or wheat smut might catch fire on threshing machines, while the machine's steam boilers could explode.¹¹

After work, the harvest "stiffs" found the food and lodgings less than luxurious. The food commonly consisted of the "four B's": bacon, beans, biscuits, and bullgravy, and often came burnt from the frying pan. The alkaline water of northwestern Oklahoma and southern Kansas could cause diarrhea or dysentery. The workers slept either on the ground or on straw piles and had to provide their own blankets, said William Casebolt, an IWW job delegate who joined the Oklahoma harvest during World War I. As farmers' incomes improved, some provided tents and a few erected bunkhouses, but as late as 1921 about 30 percent of the workers still slept in barns or granaries, often without blankets.¹²

When the work ended, the harvest hand had no guarantee he would be paid. Some farmers used pick handles to drive off workers who demanded their pay, and the migrants lacked any legal recourse to claim their wages. Once paid, a worker could still fall victim to hi-jacks or might find himself arrested for vagrancy if he failed to leave town fast enough. Fines and court costs then gobbled up the remainder of his earnings. Wobbly Nels Peterson, arrested for vagrancy in Kansas after leaving the Oklahoma harvest, saw twenty-five of the forty dollars he earned go to pay his fine. Even if a hand were not arrested or robbed, he would find that local businesses raised their prices during the harvest, reducing the value of the already minimal wages.¹³

Such treatment only strengthened the IWW's contention that the farmers, as employers, were essentially exploiters tied to the capitalist system. The Wobblies disagreed with the Oklahoma Socialist Party, which held that farmers could become a revolutionary class. Ethel Carpenter, an Oklahoma City IWW member, argued, for example, that the isolated farmers lacked the motivations and feelings of the working class and even exploited their own families. "During harvest it is easy to see the class lines drawn and if anyone doubts that the farmer makes as good a slave driver as exists under capitalism, get a job next summer in the harvest field," she wrote in the *Industrial Worker*. Like many other union members, Carpenter maintained that the IWW would find success only when capitalism consolidated the small farms and reduced the farmers to agricultural laborers. Still, Carpenter urged the IWW to organize the harvest hands because it might force the smaller farmers out of business more quickly. Other Wobblies agreed with Carpenter's position and repeatedly rejected efforts to recruit farmers, such as Covington Hall's 1913 call to include Louisiana tenant farmers in the wake of a Brotherhood of Timber Workers strike.[14]

The IWW soon came to believe that the terrible working conditions, low wages, and the poor treatment from farmers would create some class consciousness in the migratories. But the very nature of the migratory lifestyle argued against this. Rather than the franc tireurs of the revolution, the hoboes were a mixed lot.

Migratory workers were known by several names: bindlestiffs, casuals, floaters, tramps, bums, sons of rest, and, most commonly, hoboes, a term probably derived from "hoe boy." Among the migratories themselves, each term had a different meaning: a hobo was a migrant willing to work, while a tramp had given up finding work but not the traveling spirit. A bum had simply surrendered to his fate and rarely left the city. These distinctions were lost on most people, who accepted a popular image of tramps and hoboes as shiftless, dissolute, and indolent beggars prone to petty theft and drunkenness. Such views probably stemmed from a middle-class assessment of poverty and failure as punishment for sin, but newspaper articles and folktales portraying the hobo as dangerous and even murderous reinforced the stereotypes.[15]

Oklahomans, who generally saw migratories only at harvest time, clearly shared such ideas. The *Daily Oklahoman* in 1914 described incoming harvest hands as "sons of rest"—a vaudeville term—and suggested good food could lure them to work.

Farmers, the paper said, sought out the "cosmopolitan hobo": "[T]he one who has a clear eye and whose nose is not painted with a winter's dissipation." The *Oklahoman* labeled the others, especially those seeking better wages, as aging "bums" or "mercenaries" who really did not want to work. Similarly, when stranded hoboes began door-to-door begging in Enid before the 1916 harvest, the *Enid Daily Eagle* claimed the men would probably refuse work even if it were offered. When not editorializing against hoboes, local papers ran unflattering news items—articles about a railroad policeman shot by a hobo in Kansas and a farmer allegedly murdered by a tramp near Guthrie, for example.[16]

Despite the stereotypes, the majority of migratories were neither dissolute nor congenital beggars, but neither were they so radically inclined as the IWW believed. In fact, they hardly constituted a truly homogeneous class of workers at all. The vast majority were unmarried white men in the "prime of life," that is, from twenty to forty years of age. Many had either lost skills to mechanization or never possessed skills at all, which left them able to find only manual labor. About three-fourths of the harvesters were native-born whites from Oklahoma, Arkansas, Missouri, Illinois, and Kansas, while most foreigners came from the British Isles or Canada. African-American harvesters were rare outside of Kansas, and Hispanics—such as those encountered by IWW member Nils Hanson outside Enid in 1917—composed but a small fraction of the workforce.[17]

The largest portion were true "floaters" who made the harvest part of a yearly circuit of seasonal jobs. Most saw tramping as a premarital work strategy, a stepping stone to a better life. In an early study of hoboes in 1891, sociologist John J. McCook showed that most of these men either held jobs or actively sought work. Many took pride in job skills that mechanization had made redundant. The remainder were farmers, tenants, or farm laborers seeking to supplement their incomes; industrial workers laid off by seasonal fluctuations; and college students out for adventure.[18]

Attempts to organize these workers had usually failed. While some migratories showed an interest in unionism, others were openly hostile. Industrial workers often belonged to AFL unions, but those unions opposed the IWW and rarely cooperated with the Wobblies. Many of the farmers, especially the tenants, probably sympathized with the IWW, even though it rejected them as members. It seems clear that at least some of the tenants who later formed the Working Class Union had been exposed to the Wobblies' ideology while on the harvest. The college students, however, were often contemptuous of workers and labor unions. Another complication stemmed from the fact that many hoboes maintained a sort of "rugged individualist" philosophy. These men concerned themselves only with saving a cash "grub stake" to survive the winter and rejected any sort of collective action to improve conditions. Such men also refused to see themselves as displaced workers in an industrial society and often embraced the same middle-class values of the very townspeople who despised them. These were the workers the IWW derisively called "scissorbills."[19]

Given such factors, the IWW had set a difficult task for itself in recruiting migratories. But the Wobblies had already dealt successfully with other unskilled workers, overcoming obstacles such as the unskilled workers' inability to afford initiation fees and dues and seasonal job markets not conducive to stable union locals. Such factors had created serious problems for the Knights of Labor and had convinced the American Federation of Labor not even to try organizing the unskilled, although the Oklahoma State Federation of Labor did regularly endorse efforts made through the AFL's Federal Labor Union locals.[20]

But organizing migratories posed other and different problems. The urban industrial workers could draw on the sympathy of the community in which they lived and often had the support of local merchants, but the migratories were outsiders and lacked such allies. Indeed, the residents of Wheat Belt towns viewed the outsiders as criminal and moral threats to the community, attitudes reinforced by the migratory fringe of alcoholics, drug abusers, gamblers, and hi-jacks, as well as the presence of casual homosexuality among some migrants.[21]

Adding to the townspeople's distress was the common perception that a womanless, "bachelor society" such as that of the migratories was inherently violent. There is some degree of truth to this, although it probably has more to do with economic status rather than gender differences. Trying to sell their labor in a hostile marketplace, the migratories had fewer economic options than sedentary industrial workers and were therefore less likely to find security or financial success. Because they also lacked stable home and community ties, the migrants increasingly lost control over their owns lives and found themselves excluded from society at large. It is likely that this environment created feelings of weakness and uncertainty, and, rather than leading to class consciousness and solidarity, tended to produce class and ethnic antagonisms that the laborers vented on themselves and on co-workers. It also encouraged antisocial behavior such as excessive drinking, visiting prostitutes, and fighting. In Oklahoma, the state's Prohibition laws exacerbated the problem by eliminating the local saloon and thus bringing the migrants into contact with bootleggers and other criminals.[22]

A culture based on brawling and boozing clearly challenged existing Victorian values of hard work, thrift, sobriety, and family. Most townspeople therefore saw the migrants not as workers pushed to the periphery by economic forces beyond their control but as uncivilized men prone to criminality and perversion. It was in this spirit that the *Tulsa World* saw in the migrants the "survival of primitive instincts of brute force" and the potential for becoming "a lawless mob that threatened to take what they were not given."[23]

Faced with such fears, Wheat Belt authorities and business interests supported vagrancy laws that in effect criminalized unemployment. Similarly, these groups feared the IWW not just for its revolutionary rhetoric, but because they believed the Wobblies were trying to organize and make more efficient a criminal class. "The I.W.W. is pronounced one of the most vicious labor organizations in existence because of the class of its membership," the editor of the *Enid Daily Eagle* noted. So the IWW represented not just a challenge to established economic values but also a threat to social values, a mob of the unemployed more militant than Coxey's Army in the 1890s.[24]

The IWW, for its part, was fully aware of this perception and actually shared much of the middle class's antipathy toward the fringe elements. AWO member Tom Conner warned early on that Oklahoma's state prohibition law might encourage the presence of bootleggers and criminality. To counter public fears, the IWW encouraged sobriety and self-control. Delegates made an effort to bathe frequently and to "boil up"—wash—their clothes often so they would always appear presentable to the respectable elements of society, as well as to eliminate lice. Members policed their own jungles and boxcars and threw criminals as well as nonmembers from the trains. IWW papers regularly published the names of known gamblers, bootleggers, and thieves, and more than one Wobbly lost his life battling those petty crooks and gunmen. In addition, the Wobblies criticized homosexuality as less than manly and, in the case of man-boy relationships, as exploitative.[25]

While these tactics helped make the harvest safer and probably made the IWW more appealing to the harvest stiffs, Wobblies were still perceived as criminal by the public, which remained fearful of the IWW's radical rhetoric. This only frustrated the union members, who grew angry at being accused of not wanting to work and of having newspapers and employers declare that IWW stood for "I Won't Work" or "I Want Whiskey." As an editorial in *Solidarity* noted, "But GAMBLERS don't work and they are OK. Stools don't work either. And the only reason the employers are against the I.W.W.'s is that they do work."[26]

The Wobblies also came into conflict with aspects of migratory life. Both their revolutionary rhetoric and their unionism clashed with the nascent middle-class values to which many harvesters still clung. The union's class-conscious support of sobriety and opposition to gambling were in direct opposition to behaviors that—while antisocial—were defining elements of migratory culture.[27]

On the surface the IWW might understandably have given up any hope of trying to enlist the migratory worker in the cause of industrial unionism. The few outbreaks of labor unrest among harvesters as early as the 1870s usually involved local grievances, while a 1903 attempt among harvesters in Oklahoma and Kansas to gain a 20-percent wage increase came to naught. Indeed, the IWW itself at first generally ignored the harvest hands, although its 1906 convention did approve a resolution calling for the organization of the "farm wage slaves" and the use of lumberjacks as an "organizing wedge." By 1910, some factions within the IWW did try organizing the harvest stiffs, and some migratories formed IWW locals in Washington State, gaining minor wage concessions. That convinced members of the Saint Louis local in 1911 to try organizing the Wheat Belt.[28]

But *Solidarity*, the official IWW newspaper, judged the attempts futile, noting that job sharks sold phony jobs with claims of high pay to unsuspecting hoboes. "Promises of $3.00 per day and steady work for hands will materialize at $1.50 and $2.00 when it don't rain," the paper soberly noted. Its prediction proved accurate. By August 1913, the IWW had succeeded in organizing three locals of agricultural workers, two of which had already disbanded. Similarly, of thirty mixed locals that included agricultural workers, only eleven remained. The spring 1914 campaign of Kansas City Local 61 to raise the farm wage to $4.00 a day also ended unsuccessfully, though in some areas daily wages rose to $2.50 and $3.00.[29]

But the Kansas City effort did teach the IWW that organizing the Wheat Belt was more complicated than organizing a factory. Because few workers actually lived in the Wheat Belt, organizers could not rely on community, family, or ethnic ties, as they had in the union's East Coast strikes. Recruiting the hoboes would require a strong, coordinated effort to attract new members and show them how to use job actions such as slowdowns and brief strikes. It would have been difficult work for any labor union, let alone one as weak as the IWW at that time. But the surprising incompetence of state and federal employment bureaus in regulating the harvest hiring process gave the IWW an opportunity it did not waste.

The need for harvest workers led Oklahoma Territory officials in 1905 to join with Iowa, Kansas, Missouri, Nebraska, and South Dakota to form a Western Association of Free State Employment Bureaus to supply harvesters. Coordinating the labor supply over a large area proved unwieldy, and the association collapsed by 1908, forcing Oklahoma to rely on its own state employment bureaus and, later, on the new federal Department of Labor.[30]

Like scientific management in factories, the agricultural employment bureaus were part of the Progressive Era's impulse to rationalize the labor force. But the bureaus were not suited to the farmers' more decentralized, seasonal needs. Instead of rationalizing the labor force, the employment bureaus created labor surpluses, which forced wages down. This led to large, roving bands of the unemployed throughout the Wheat Belt, producing a summer-long disaster in 1914.[31]

In May 1914, midwestern farmers, fearing IWW activity like that in California and the Northwest, worked to establish the National Farm Labor Exchange. The new organization met with the Bureau of Labor's Division of Information in Kansas City to set up a labor recruitment campaign for the 1914 harvest. Handbills, posters, and newspaper advertisements appeared nationwide calling for 100,000 men and optimistically promising three to six months of work at $2.50 to $3.00 a day.[32]

But the National Farm Labor Exchange failed both to limit the number of workers and to seek men fitted to the work, while the state employment bureaus failed to distribute properly the flood of workers who came forward. Some harvest towns overflowed with eager hands, while others were starved for help. The total inadequacy of the plan emerged with the opening of the Oklahoma harvest. Although heavy rains and flooding had delayed the start of the harvest, farmers at Woodward and Alva still needed 2,500 men in late June, when the harvest should have been ending. The farmers also claimed that 14,000 men were available, but the railroads refused to give them free passage. Railroad agents added to the confusion by sending many workers to Oklahoma City, which lay outside the harvest belt. In addition, job sharks had sold bogus names of employers to many potential workers, and the towns treated all the migratories as vagrants because so many men were forced to sleep in parks when harvest jobs fell through. Many Oklahoma towns, fearing trouble, tripled their police forces. The threshing season proved even worse, with only one-fifth of the migratories needed for the work.[33]

Some farmers remedied the situation by stopping trains and pulling men from the boxcars, thereby denying help to other communities. Businessmen at Cherokee started a car service to collect harvesters from other towns and get them to farmers. Despite these measures, Woodward farmers still needed an additional 2,000 men to complete the harvest. Saying the state faced a calamity, the *Daily Oklahoman* urged the use of federal funds to move the workers before the crops were ruined. But the mass of men proved unmanageable, and by June 17, the state employment bureau began sending incoming harvesters directly to Kansas and suggested that workers already in Oklahoma go directly to the farmers.[34]

The labor glut forced wages in some areas of Oklahoma and Kansas to drop to 50¢ a day, but men still could not find work. In northern Kansas, workers were so scarce that farmers were offering $4.50 a day yet failed to attract help. Reports circulated of angry harvest hands commandeering freight trains and forming informal unions to demand better wages by withholding work. Such activity frightened state officials, who sought scapegoats for their own failures. The mere presence of Wobblies at Enid led Assistant Labor Commissioner W. G. Ashton to blame the IWW for wage demands. The *Daily Oklahoman* noted none of "that kind" were at Woodward, although the labor-starved community had few harvesters of any kind.

The confusion initially worked to the Wobblies' advantage. The IWW press committee at Enid claimed delegates had organized 300 men, all of whom voted to fight for $2.50 and the ten-hour day. A Wobbly-organized march through Enid on June 4 ended with area residents and local businesses raising $22.50 to feed the harvest stiffs. Commissioner Ashton met with an IWW committee on June 6 and urged the men to accept the minimum wage. But the IWW organizers assessed their gains too optimistically and confused the area's need for hands with support for organized labor. "This town is strong for the one big union and are doing all in their power to get our demands granted," one organizer said. Later harvests revealed a more hostile populace.[35]

The IWW did make some gains in the 1914 harvest. In parts of Kansas they helped raise wages to three dollars a day and won the ten-hour day. They also made IWW policies known to a wider number of workers. But the chaotic harvest also revealed how woefully inadequate the IWW's recruiting strategy was. The union's ninth convention that fall responded by endorsing a proposal from Frank Little calling for an April 1915 conference in Kansas City to develop a strategy to recruit the hoboes.[36]

In adopting Little's proposal, the IWW began emphasizing the recruitment of western migratory workers at the expense of the East Coast unskilled factory workers. The process led to increased centralization of control by the general headquarters, which later contributed to the union's essentially fatal 1924 split. While locals previously operated separately, they were few and far between in the Wheat Belt, and such a widespread campaign required a national union: the Agricultural Workers Organization no. 400. Due to the scarcity of locals, the national secretary issued blank cards and dues stamps to members who could sign up interested workers wherever they happened to be. This practice created the "job delegate" system, a hallmark of IWW recruitment. Under this system, members set wage and hour demands beforehand, selected an individual or committee to negotiate with a farmer, then, as a group, ratified any agreement. The goals for 1915 were the ten-hour day, a three-dollar minimum wage, overtime pay, good board, and clean beds throughout the Wheat Belt—all improvement that were easier to obtain by group rather than individual negotiation. Besides emphasizing solidarity, the system also allowed workers to keep in touch both with each other and with the organization. To encourage recruitment and prevent distraction, the AWO elected to drop propagandizing for the revolution and instead to emphasize dues and memberships and immediate job-related goals. *Solidarity* warned the wheat growers of Oklahoma and Kansas, "The above demands are asked of you, and if granted, satisfactory work will be done."[37]

While some Wobblies saw the job delegate system as a means to ensure worker control of the IWW, they also knew it had a serious weakness: the tendency for the delegates to focus their efforts on the jungles. While the delegate could reach more men this way, it also meant that nonunion workers actually got the jobs and the IWW could exert no pressure on the job itself. Before the 1915 harvest E. W. Latchem urged the delegates to abandon the jungles and recruit only on the job. "The I.W.W. may not be able to do much this season except for a few localities, but we will have a good nucleus and better sentiment for next year's work," he wrote in *Solidarity*.[38]

But despite cautions from Latchem and the editors of *Solidarity*, AWO job delegates optimistically proceeded with plans for the 1915 harvest. State officials expected a large harvest, with prices high as a result of the war in Europe. The IWW hoped to gain a share of the anticipated profits for the workers. By May 1 the AWO had prepared a virtual army of delegates for the initial drive at Enid. Several of the organizers had Oklahoma ties: Latchem had spent part of his youth in Oklahoma, and W. C. King had organized oil workers at Drumright, while George King and Jack Law had organized an oil workers' local at Tulsa in the spring of 1914. With the beginning of the state harvest set for June 8, *Solidarity* urged the delegates to line up workers quickly at the two-dollar initiation fee and dues of fifty cents a month.[39]

Arriving in the fields, the Wobblies gathered outside town and established a jungle, preferably near a stream. The job delegates called a meeting and elected committees to keep the camp clean and solicit work in town, with any pay going into a common fund. An elected treasurer kept tabs on the accounts, while other committees, such as the "spud and gump" brigade, foraged or begged for food and supplies and did the cooking. The idea was to create in microcosm the workers' society the IWW would bring about by revolutionary action. But, as Latchem feared, this focused efforts on the jungles and not on the job. Also, while the AWO convention abandoned streetspeaking and soapboxing as ineffective, Walter Nef, the AWO secretary, arrived in Tulsa on May 6 to find three job delegates spending all their time doing exactly that.[40]

Nonetheless, fifteen IWW job delegates arrived at Enid on May 30 and established a jungle at the south end of town. Unfortunately, they quickly learned the state employment bureau had warned harvesters to avoid both the towns and union organizers, whom officials claimed would just fleece the laborers and halt the harvest. The job delegates took the problem in stride, arguing it could mean higher wages as a way to deter union organizers. Instead, the Wobblies said this would allow them to focus more attention on obtaining the ten-hour day, clean board and beds, no discrimination against union men, and a minimum wage.[41]

Despite the delegates' optimism, the state and federal free employment services complicated matters by flooding the job market. An estimated 80,000 job seekers were stranded in Kansas. The Oklahoma employment agency refused to send IWW members or suspected members to jobs and required hands to have references for work. Those who asked about wages failed to receive a job ticket. While the Farmers' Educational and Cooperative Union had agreed to fix wages at

$2.00 to $3.00 a day, the actual rate was about $1.50 for haying and $2.50 at best for the harvest. Many had paid up to $50.00 for transportation to Oklahoma, despite claims of free rail transport. With wages low and jobs scare, some harvesters at Enid were reduced to cooking potato peelings, and others took low-paying railroad section jobs. Railroads ordered their brakemen to keep IWW members off the trains, while Wobblies faced the usual jungle dangers, with gamblers killing one member in Kansas.[42]

When the employment bureau tactics proved inadequate, authorities simply began arresting suspected IWW members and sympathizers. In late May, the city of Enid had opened a "hobo hotel" in an abandoned garage, euphemistically named the "Hotel de Gink," in order to end street begging and to weed out agitators. Police directed harvesters to the "hotel" and away from the jungles and urged housewives to refuse hoboes' requests for handouts. The experiment ended when officials discovered the IWW found the hotel a perfect location for organizing.[43]

On May 31, Enid police, Garfield County deputies, and local militia raided the hotel, arrested three hundred men—including several IWW members—and forced them onto outgoing freight trains. Police also jailed one IWW member for arguing in a city park, while three other Wobblies faced vagrancy charges. The mayor of Enid blamed the hoboes themselves for the arrests, claiming they had pressured local merchants for food and supplies, and he also announced an end to a city program that provided two daily meals to early arrivals. City officials and local newspapers charged the IWW with causing the city's troubles and claimed professional tramps had burned the doors of freight cars.[44]

The Oklahoma experience convinced the AWO leadership that a strategy such as Latchem's was necessary. The union's national office told members to deny that they belonged to the IWW until they were on the job and replacement workers were scarce. One tactic was to rip up their union cards in front of potential employers—the cards could be replaced later. Once on the job, the delegates were to demand a three-dollar daily wage and begin a slowdown if refused. The leadership believed this so-called strike on the job would prove effective because most farmers needed the work done before the heads of grain shattered and would be unable to find replacements soon enough. Should the farmer fire them anyway, the workers were to demand their wages plus the fare to town.[45]

Because it was impossible to put up a true picket line, the Wobblies turned to direct action to enforce their demands. Active resistance in towns such as Enid convinced the AWO to put the same tactics into full use in the Kansas harvest. As *Solidarity* informed its readers, "$3.00 a day—shocks right side up; $2.00 a day—shocks upside down." The changes allowed the AWO to entrench itself in the fields and to raise wages to $3.50 and $4.00 a day in some areas. From July to the end of the year, AWO delegates signed up 2,280 new members, putting $14,133.00 in the union's treasury. The funds enabled the AWO to open a permanent union office in Minneapolis and to finance a West Coast harvest campaign. The money also fueled a resurgence of the entire IWW, funding the rental of offices in Chicago and of a printing plant.[46]

Buoyed by its success, the AWO immediately prepared for the 1916 harvest. Delegates were to demand $4 for the ten-hour day, with 50¢ for each overtime hour, good board, and clean sleeping areas with fresh bedding. The AWO also boldly demanded that all hiring of hands be done through the IWW hall or through job delegates, an unrealistic goal as the union had few halls in the Wheat Belt. To aid recruiting, the union placed stationary delegates in key towns, especially Enid and the Kansas towns of Wichita and Ellis, but over three hundred job delegates would do the actual work of distributing union literature, holding recruitment meetings, and signing up members.[47]

The AWO expected war. When the agitation committee in late 1915 learned of a plan to import 30,000 southern blacks for the harvest, it declared it would use African-American organizers to recruit the men. While that plan never came to fruition, others did. A February 1916 farm labor conference in Kansas City announced it would eliminate "riff raff" and better coordinate the harvest labor supply. W. G. Ashton, now Oklahoma labor commissioner, said the state would also avoid commenting on the need for harvesters to guard against the return of stick-up men and troublemakers such as the IWW.[48]

The labor commission worked with the state Farmers' Union to set a fixed wage of $2.50 a day. The farmers hoped organizers would leave town once they saw they could not raise the wage. The tactic seems to have worked initially, as the AWO enrolled only seventy-five new members at Enid. But the low enrollment may have reflected more a lack of funds by harvesters rather than a lack of union sentiment. Delegates said many of the migratories expressed interest in the AWO, a fact they believed was borne out by the complaints of farmers at Cherokee that hands demanded chicken three times a day and cherry pie.[49]

A more critical problem for the AWO was the impracticality of trying to control hiring. Farmers still met prospective hands at the railroad depots and engaged in curbstone bargaining. But even if they could not directly control hiring, the increase in wages to $2.50 in Oklahoma does suggest the IWW was influencing the labor market. Newspapers in the Grain Belt conceded that wages were higher but attributed it to the war in Europe placing a premium on agricultural products. "The ripening wheat is being put in the shock and stack without prices being dictated by agitators and disturbers," the *Daily Oklahoman* declared. That failed to stop the paper from blaming the IWW when hands did demand better pay and shorter hours.[50]

The failed Oklahoma campaign forced the AWO to alter its tactics again. Delegates now received a fifty-cent commission for each new recruit, and members used strong-arm tactics to eject criminals from the hobo jungles and freight trains and to coerce nonmembers into joining the union. One common strategem was the use of "flying squads," groups of union members who moved from location to location to enforce compliance with the AWO. The tactic drove away most criminals as well as those workers who refused to join the AWO and effectively gave the union control over hiring as only card-carrying union men were allowed on the boxcars.[51]

In addition, the Wobblies rejected passive resistance and fought back when they encountered opposition like that in Enid in 1915. Members threatened to storm jails to free comrades and organized boycotts of local merchants who overcharged harvest hands. Sabotage increased as well, but not to the extent reported in the press. Wobbly Jack Miller said that if hands were pressed by a farmer to work faster, they might feed bundles into a thresher too fast so it would clog; but they would not permit wanton destruction, such as burning crops. "If the farmer is disabled, the job stops," he recalled. "We were there to work and earn wages, not to destroy."[52]

The new tactics helped increase AWO membership to 20,000, or nearly one-third of the total IWW membership. The dues and fees also brought $49,115.00 to the IWW treasury. Still, some IWW members feared the AWO's approach only added to the union's negative image. They sought a more "scientific" emphasis on combating local commercial clubs and employment bureaus and believed the fifty-cent commission caused delegates to sign up too many questionable recruits. Members in other industrial unions worried that the AWO would send delegates into their industries and contended that it was becoming too independent. The fact that the AWO controlled more delegates than other unions at the 1916 IWW convention did not help matters.[53]

Any hope of changing the tactics ended in November 1916, when Bill Haywood and the main IWW leadership wrested control of the AWO from Walter Nef in a dispute over the agricultural union's power. Nef's replacement was Forrest Edwards, a Seattle native who in 1914 had served as secretary of the IWW oil workers' Drumright local. Edwards approved of the strong-arm approach and chastised its critics as old-fashioned. "This new blood is putting over stuff and getting away with it so that the old Wobbly seems amazed at it," he wrote. Still, with the success of the 1916 campaign, the AWO had revitalized the IWW, and union leaders looked with hope toward the future and the start of the Oklahoma harvest in Enid in 1917.[54]

The Oklahoma harvest campaigns marked a major change in the IWW. With the decision to recruit migratories, the union leadership broke with its previous membership of unskilled workers in mass production industries. The decision had both positive and negative effects on the union, increasing the IWW's financial and membership strength but leaving it more vulnerable to attacks from the business community and obsolescence due to technological change.

With its campaigns in 1915 and 1916, the IWW moved away from revolutionary rhetoric and, as did the AFL, toward more conservative issues such as better wages, food and lodging, and working conditions. The need to deal effectively with these issues helped create the IWW's trademark recruiting technique of roving job delegates. Because of its location, Oklahoma—and Enid, in particular—became a testing ground for organizing tactics. The AWO also used the wage achieved in Oklahoma as a benchmark for other states. If the state's wages rose, as they did in 1916, it meant better pay in the rest of the Wheat Belt. But to achieve these goals the AWO turned more toward direct action and sabotage, which only added to the popular image of the IWW as violent.

The decision to recruit harvesters brought the IWW financial stability and a larger membership, although at the price of greater centralized authority within the union and the acceptance of the erroneous idea that the migratories were the vanguard of the revolution. In reality, the floaters proved far more conservative and more like the middle-class townspeople who feared them. In addition, rather than being the product of the rationalization and mechanization of large-scale agriculture, the migratories were actually part of a transitional phase in the process. As a result, the IWW's position would grow precarious, caught between the need for dues and fees from the migratories and dependence on a group of workers whose way of life would soon disappear. At the time the IWW's success in the Wheat Belt outweighed such concerns and allowed the union to move into other industries, especially the Oklahoma petroleum industry.

3

Organizing "Oily Willy"
The IWW in the Midcontinent Oil Fields

"THE I.W.W. has got the clearest and cleanest cut declaration of any outfit going," wrote Cushing oil worker James Koen to an Arkansas friend in 1916. "I know the organization has got no halfway grounds with the capitalists." Branch secretary of a Cushing IWW local, Koen believed the union had just the program for Oklahoma's oil workers. In a letter to Bill Haywood, he urged the IWW to organize aggressively in the Midcontinent oil fields of northern Texas, Oklahoma, and southern Kansas.[1]

The AWO expressed interest in organizing the oil fields. Oil workers often joined the harvest when work was slack, and a growing number of harvest migratories found jobs as pipeliners and ditchers during the fall and winter. The union's leadership realized the job delegate system could successfully recruit these workers. In addition, this offered the IWW a chance to expand its base and operate more effectively as a labor union. The renewed oil boom in Oklahoma—caused in part by the start of World War I—further convinced the IWW leadership that the industry was ripe for unionization.[2]

Next to the harvest campaigns, the Midcontinent campaign represents the IWW's major organizing effort during World War I. On the surface, organizing should have been simple. The workers were isolated and performed relatively dangerous tasks, factors that should have produced radical attitudes. But as Wobblies learned, the oil workers proved less receptive to unionization than had the migratories.

Much of this may have had to do with the high wages oil workers earned, as well as the instability of oil-boom towns, which discouraged worker solidarity. Controlled by developers directly tied to national oil companies, the boom towns exploited workers with high prices, illegal liquor and drugs, gambling, and prostitution. Authorities in such towns were usually corrupt and often criminals themselves. As was true of harvests, a large pool of willing, often desperate, men flooded the oil fields seeking a limited number of jobs. Despite such obstacles, the IWW made some progress in organizing unskilled hands such as pipeliners. Using tactics from the harvest and lumberjack campaigns, job delegates focused on workers in isolated camps, away from the boom towns; downplayed revolutionary rhetoric; and utilized short strikes to achieve workplace goals.

But unlike the AWO, organizers for the new Oil Workers Industrial Union (OWIU) found themselves dealing not with individual farmers for whom every lost second could mean economic disaster but with vast national corporations that could draw on large financial resources to outlast strikes and enlist the repressive power of the state and federal governments. The mere presence of the Oil Workers Industrial Union created fear among business executives and government officials, especially during World War I, when oil production increased to meet wartime demands. Facing labor unrest and charges of war profiteering, the oil companies found it convenient to claim the IWW posed a direct threat to the war effort. The wartime need for petroleum also gave the federal government the means to begin extensive prosecution of the IWW, culminating in the infamous 1918 Chicago conspiracy trial of the union leadership.

That Oklahoma was the focus of federal concern over the IWW should be no surprise. At the time, the state was the largest producer of petroleum in the nation. The majority of the oil came from the huge Midcontinent field, which was 75 miles wide, 175 miles long, and contained 30,000 wells. Included in the field were the Glen Pool near Tulsa, the Cushing Field in Creek County (which provided half the Midcontinent's output), and the Healdton Pool. The Oklahoma fields, especially the Glen Pool, proved ideal for development. They were shallow, making drilling expenses low, and had plenty of natural gas to power the pumps. The oil was "sweet": light-bodied, relatively free of sulfur, and with a paraffin base useful for producing kerosene, then the mainstay of the oil trade. By 1916 the Midcontinent alone produced nearly 291 million barrels of oil, worth more than $139 million, or more all than all other regions.[3]

While small independents did most of the drilling, the major national producers such as Royal Dutch Shell, Sun Oil, and Gulf Oil effectively owned the oil because they owned the pipelines and the storage facilities. By far the biggest players were Standard Oil, of Indiana, a spinoff of the breakup of the Standard Oil trust in 1913, which owned Carter Oil Company, a major controller of Midcontinent real estate; and Prairie Oil and Pipeline Company, the principal pipeline company of the region, with 900 miles of main (or trunk) lines and 2,000 miles of smaller gathering lines.[4]

Standard of Indiana also controlled three of the region's five major pipelines: Prairie Oil owned two, one of which was a seven-hundred-mile trunk line from Cushing to Standard's refinery in Whiting, Indiana, near Chicago; while a third was owned by Magnolia Oil, an independent that was in fact controlled by executives of Standard Oil of New Jersey and Standard Oil of New York. The remaining two lines belonged to the Texas Company and to Gulf Oil. All five lines were valued at $43 million. At a cost of $9,000 a mile, lines were expensive to build, so only the major oil firms could afford it. They therefore decided when to build a line, when a field had produced enough to justify connecting it, being a monopoly, and what price to pay for the oil.[5]

Besides controlling the market, the firms wielded considerable influence over the Oklahoma legislature and the press. In fact, the original Standard Oil trust had twice bribed the Oklahoma territorial legislature to stop attempts to regulate the quality of its kerosene, while former Oklahoma governor

Charles Haskell had once tried to buy off Ohio's attorney general to keep him from pursuing a case against the petroleum trust. It was also common for oil companies to subsidize local newspapers. It was not for nothing that Upton Sinclair had remarked, "In Oklahoma, everything is Standard Oil." In electing to recruit oil workers, the IWW had picked some powerful opponents indeed.[6]

Still, the Wobblies remained convinced they could do it. All they needed to do was look at the oil-boom towns to find conditions with which they were already familiar from their experiences in the western mining and lumber camps. Like those regions, the oil-boom towns grew up around readily accessible resources—in this case, shallow oil pools that attracted entrepreneurs with little capital. The developers then either built new towns or expanded existing ones. Towns located in large fields or at the junctures of large fields were more stable than those in smaller fields, but all shared traits with western mining towns: a huge influx of generally itinerant people, poor-quality housing, and lots of easy money, which attracted the "vice" industries. Prostitution, illegal liquor, and gambling were common. Kiefer had its own "Bowery," with brothels and "bobcat" whiskey joints, while Drumright possessed a notorious brothel called The Hump. Tulsa reportedly had one hundred places that sold illegal liquor, and at least twenty-two Tulsa hotels harbored prostitutes. They often worked directly on the line and were usually tolerated if they did not solicit in the open. Illegal saloons offered whiskey and "Choctaw beer," but druggists also openly sold alcohol in "medicinal" form or as "Jamaica ginger" and bay rum.[7]

Law enforcement was lax, at best, as local officials often were themselves ex-convicts or had a financial interest in the illegal activities. For example, Bud Ballew, the sheriff of Ardmore in 1921, was part owner of the brothels in Healdton, while Charles King, the Ringling police chief, was part of a whiskey-smuggling ring based in Wichita Falls, Texas. Occasionally such officials faced corruption charges, but few were ever convicted. Besides casting a blind eye on vice, the law also seemed unable to stop other crimes. Robbery was common as employers often paid wages daily in cash. Drilling crews working at night were usually the victims.[8]

Crime and vice were bad enough, but living conditions were deplorable. Typical of the boom towns was Kiefer, in Creek County, where waste oil ran down unpaved streets that became oceans of mud after a heavy rain. Housing usually consisted of tents or wooden shanties that lacked indoor privies and running water. The poorly built structures did little to hold back cold weather, and winter often produced high death rates among children. Some families also lived in boarding houses, crowded into one or two rooms.[9]

Single men fared little better, usually staying in cheaply built hotels known as flop houses. Proprietors charged workers fifty cents for eight hours to sleep in beds placed in rows with an aisle down the center. Workers secured their own valuables, usually placing the cot legs in their valued steel-toed work boots to deter theft. But even this poor housing was in short supply, and some workers slept in the open along creek banks.[10]

Despite the conditions, job seekers attracted by advertisements and flyers filled the towns, the capital-intensive oil industry required only a small labor force and could not absorb all the applicants. About 20,000 oil workers, at most, found jobs in Oklahoma, and the rest made up a huge labor surplus. One 1914 source claimed Tulsa had 15,000 unemployed men looking for oil field work, and that at a time when the city's total population was only 35,000. While the Tulsa claim may be exaggerated, unemployment always increased once a pool was opened and the oil began to flow. This forced workers either to move on to new fields or hope for work laying or maintaining pipelines.[11]

In many respects oil workers resembled their harvest counterparts, but some important differences existed. The first to arrive in a new field were the "boomers": rig-builders, drillers, tool-dressers, storage tank builders, and some pipeliners—who were paid a premium wage of six to seven dollars for a twelve-hour shift. These men were generally younger and unmarried, as the oil companies believed single men could endure harsher working conditions. Later came the "boll weevils," unskilled farmers or farm laborers hoping to supplement their incomes, especially during downturns in the agricultural economy. Most became lease workers or "roustabouts," a generalized term for common laborers. Unlike the boomers, many were married and could return to the farm when work there became available. A third group were the "drifters," often older men, some ex-convicts, others outcast because of their homosexuality, still others dogged by problems such as alcoholism or drug abuse. Rarely making good hands, the majority ended up in oil work by chance rather than choice.[12]

The shifting nature of oil work forced all three groups to be migratories. But unlike the harvest hands, who traveled a wide circuit from the Grain Belt to the West Coast, oil workers limited their migration pattern to Oklahoma, Texas, and oil fields in neighboring states. A few, like Wobbly James Foy, followed a wider path. Foy did pipeline work in Newkirk for the Oklahoma Pipeline Company in early 1923, then took various jobs in Idaho, Washington, and California. Oil workers recognized a commonality with the harvester and readily called themselves hoboes or "high-class tramps."[13]

The working conditions should have aided the IWW in recruiting. Most workers generally worked a twelve-hour "tour" (pronounced "tower") seven days a week. Safety was lax in the fields, and workers often lost hands and fingers to machinery or were crushed under collapsing stacks of pipe. The nitroglycerine used to "shoot" the wells—that is, to make a basin in the drill hole to hold the oil—often exploded, as did the natural gas that drove much of the machinery. The oil industry was so fraught with danger that it accounted for 40 percent of Oklahoma's disabling industrial accidents and 25 percent of its work-related deaths.[14]

But other factors made these workers harder to organize than the harvest hands. On the one hand, drillers, tool-dressers, and rig-builders were often employed by outside contractors. Better paid, they tended to be more conservative. Some, however, were members of AFL unions, and evidence suggests the IWW's Construction Workers Industrial Union in 1917 did recruit a number of rig-builders. On the other hand, roustabouts commonly came from rural areas and, because they intended to return to the farm, often saw no need for labor organizations. Only the pipeliners, who were directly employed by oil companies or their subsidiaries, proved amenable to unions.[15]

Equally divisive was a strict hierarchy among the workers. At the top were drillers and tool-dressers, followed by tank builders and rig-builders. These men took pride in their job skills and segregated themselves from the pipeliners, roustabouts, and other semiskilled and unskilled workers. Fights between the groups were common, such as one that broke out at a camp near Drumright in October 1914 when pipeliners were told to get their lunch at the tank builders' mess.[16]

The anti-union stance of the oil companies added to the IWW's difficulties. The firms regularly employed spies to report unrest, and the larger companies maintained sizable private police forces. Employers also used the "race card," playing off the few black and Mexican workers against the whites to keep wages low. In addition, oil town businessmen proved particularly eager to form Ku Klux Klan units, and local police repeatedly clashed with workers.[17]

Even when workers were organized, their migratory nature restricted job action. The masses of unemployed willing to scab militated against most strikes. If a worker filed a complaint against an employer, it often took so long to resolve that by then the worker had already moved on to another job.[18]

For these reasons the IWW concentrated its efforts on pipeliners. Pipeline crews generally worked away from the town and lived together in mobile camps until the job was completed, similar to the pattern in Pacific Northwest lumber camps, an IWW stronghold. When OWIU organizers found support for job action, the weapon of choice was the short strike. This either immediately achieved the workers' demands or resulted in the firing of the entire crew. But even when the strikes failed, they often produced positive effects among superintendents who knew they had to improve conditions if they wanted to keep future employees.[19]

Once a field was developed, pipelines were needed to send the oil to the refineries. Crews of 100 to 150 men hauled, dragged, and even floated the huge sections pipe from railroads located as far as twenty miles from the pipeline right-of-way. While most of the equipment was hauled by horse-drawn wagons, it was not uncommon for men to carry the joints—which could weigh up to six hundred pounds—on their shoulders for the last few miles.[20]

The pipeline camps were placed well away from towns to prevent the men from getting drunk. "They didn't do it because they were big-hearted, not by a helluva lot," one pipeliner quipped. Workers lived in tents lighted by kerosene lamps and heated by natural gas from the wells. Most tents had dirt floors and little except for a few boards placed at the sides to keep out the wind.[21]

As did the harvest hands, oil workers found the food and bedding materials poor. Local water was generally unfit and often carried diseases such as typhoid. Many workers complained that the mattresses and cots were "crummy"—infested with lice and bedbugs—while ants overran everything. Fumigation was futile. A camp veteran recalled, "It was spitting on a fire to put it out." Once the work ended, the pipeliners often found their pay was in the form of a "timeslip," a check no one would cash.[22]

The work itself was demanding. Jobs began with two camps at the ends of the proposed pipeline, the crews then building toward the middle. A camp was usually set up five to six miles ahead of the job so the pipeliners worked their way up to it.

When the pipeline was five or six miles past the camp, the camp was torn down and moved ahead of the workers, starting the process all over again.[23]

"Bush crews" first cleared debris and trees from the right-of-way, then graded the surface. Following them were "stringing crews," who laid out the pipe along the right-of-way. This often proved dangerous, especially when stringing the line across flooded streams or rivers. Crews had to screw together sections of pipe and winch them across from bank to bank. More than one pipeliner drowned during these operations. Once the pipe was strung, work gangs numbering about seventy-five men began digging the ditches to hold the pipe. After the ditch was cut, the pipeliners began connecting the joints, which ranged from four to sixteen inches in diameter and from twenty to forty feet in length.[24]

Before the use of welded pipe in the 1930s, the pipe was literally screwed together. The "tong gang," or "laying crew," first raised the pipe off the ground, then swabbed it to remove dirt or any animals such as rabbits which might have nested in the line. The threads were oiled, and the collared end of the next pipe fitted to it.

After a man called a "stabber" made certain the joints were aligned, the crew used ropes to turn the joint until it became difficult. At that point teams of four to six men used heavy wrenches called tongs to tighten, or "buck up," the joint. Most tongs were large enough to require three men, and the workers "spelled off," or relieved each other, at various stages. The pace of the turning was synchronized by a "collar pecker," who beat out a rhythm on the pipe with a pair of ball-peen hammers. As the turning grew most difficult, the "collar pecker" would "knock on," or call for, more tong crews.[25]

Once the joint was tightened, the "dope gang" painted the pipe with sealant—usually a red, lead-based paint or an asphalt-based enamel—then wrapped it in either tar paper or felt and applied a second coat of sealant. Only then was the pipe lowered into the ditch and buried using shovels or a "marmon board," a five-by-three-foot blade pulled by mules that was the forerunner of the bulldozer.[26]

All of this required teamwork and cooperation. The camaraderie created most likely aided the IWW organizers' recruiting activities, as did the shared experience of the low wages and poor work-camp conditions. Combined with the pipeliners' physical isolation from towns, these conditions served to promote a sense of separation from society and of exploitation. Pipeliners, more than other oil workers, were therefore more likely to see the oil companies as autocratic and themselves as more populist.[27]

But even shared grievances and experiences were insufficient to aid unionization. For example, in February 1915, Wobbly J. M. Healy was part of a crew laying pipeline for the Texas Company from Drumright to Yale in Creek County. The area was so isolated that the men either walked eight and a half miles to find a place to sleep or slept in a nearby pump station. Yet the men showed no interest in a union. As Healy wrote in *Solidarity*: "When I asked some of them why they did not organize, the answer was they were perfectly satisfied."[28]

The IWW itself could also accept some of the blame for the oil workers' lack of interest in unions. As it had with the harvest migrants, the IWW initially ignored the oil field workers. Although William Trautmann, IWW secretary-treasurer, urged the union's mining department in 1906 to recruit oil field and refinery workers, the Wobblies made little effort to bring them into the fold. Between 1911 and 1913, the union had only fifty to sixty oil workers as members, and its only strike in the industry occurred in 1910 when Hispanic laborers sought higher wages at a southern California gas works.[29]

The failure of the AFL's organizing efforts may also have tempered the Wobblies' interest. Standard Oil had a long history of antilabor action, breaking strikes by coopers and refinery workers in the 1870s and 1880s and later forcing employees to take an oath on the Bible against union membership. By 1900, Standard had also broken the locals of the International

Brotherhood of Oil and Gas Well Workers in western Pennsylvania, Ohio, and California. In late 1905, the AFL countered by organizing locals along the Gulf Coast, which lay outside Standard's control. The first oil workers' local chartered in Oklahoma was at Kiefer in 1907. The branch had 160 members but was short-lived as workers drifted away when drilling operations stopped in the wake of the Panic of 1907. That depression also wiped out the gains in Texas and Louisiana, although those areas later become AFL strongholds.[30]

Oklahoma and southern Kansas proved more fertile for the Wobblies. Individual IWW members had worked for some time in the region, but an influx of blacklisted lumber workers from the Sabine River pine forests of eastern Texas and western Louisiana brought the union to the area in force. These men were members of the Brotherhood of Timber Workers (BTW), an IWW affiliate, which had waged bitter, often bloody, strikes against the Southern Lumber Operators Association and its member companies from 1912 to 1913 until the operators broke a 1913 strike at Merryville, Louisiana. Out-of-work BTW members came to the Oklahoma oil fields, and several BTW veterans, such as Phineas Eastman, later became important leaders of the OWIU. The BTW's direct experience with giant corporations and with the employers' use of labor surpluses to keep wages down gave its members an understanding of the similar situation in Oklahoma. As rural southerners and tenant farmers themselves, they also had an advantage in communicating with their Oklahoma counterparts forced to leave the farm to find work.[31]

By late 1913, the IWW General Executive Board came to agree with the rank-and-file members and sent organizers to Tulsa and Drumright to form oil workers' locals. Frank Little and A. W. Rockwell organized a local at Drumright, while Jack Law did superb work in organizing Tulsa's local. A veteran of several IWW free-speech fights, Law had previously organized construction workers. On December 20, 1913, with oil worker George Fenton, he formally set up Local 586 and within a month had signed up one hundred members, mostly pipeliners.[32]

The Wobblies began soapboxing on Tulsa streets, and organizers appeared in most camps. When one employer tried to hire some Bulgarians at a daily wage of $2.00 instead of the going rate of $2.25 plus board, the local sent in two Bulgarian-speaking delegates who got the men to refuse the lower pay. The local also gained a small victory by getting some pipeline projects either to cease work on Sundays and holidays or pay time-and-a-half.[33]

Unlike many of the harvest delegates, the oil field organizers clearly saw themselves as trying to establish a labor union with long-range goals. Most had experience on the pipeline crews and did yeoman work in spreading what Law called the principle of education and solidarity. Membership grew rapidly, and in mid-March the Tulsa local actually ran out of initiation and dues stamps and had to ask other locals to provide them. Adding to the rolls were several AFL-affiliated boilermakers and some members of the AFL oil and gas workers' local angered after their business agent absconded with the local's funds.[34]

By April, Local 586 had over three hundred members and was preparing a push to gain the eight-hour day, a $3.50 daily minimum wage, and a half-day's pay for laying one or more joints. They also sought pay for transportation to the work site, sanitary bedding, tents set aside for reading and writing and for bathing, and a set price for meals at the camps.[35]

The oil companies now took notice of the OWIU. One Tulsa oil man declared, "If the demands of the organization are acceded to, we might just as well get ready for others to follow and plant flowers around the derricks, put in swimming pools on the leases, and open club rooms and other forms of amusement." Others, however, believed that the oil field workers were too patriotic to affiliate with an organization of "flag-tramplers."[36]

Tulsa city officials were also worried. They viewed the Wobblies' use of street agitation and free-speech fights as subversive and seditious. They found the nonviolent nature of the protests both scary and perplexing. To deal with the IWW, the

Tulsa city commission in April 1914 passed an ordinance forbidding more than three people from congregating on a public street. Armed with the new law, Tulsa police broke up street meetings and arrested IWW members on vagrancy charges, claiming the Wobbly agitators were anarchists who spread "vicious vaporings" and called the American flag a "dirty rag."[37]

When this tactic proved ineffective, the police started arresting the speakers. In June 1914, E. W. Brink—a Wobbly and a Socialist Party member—was arrested after he made a street speech that detectives claimed was filled with "atheism and anarchy" and that did attack Christianity and the Tulsa city commission. Brink, who had already been arrested once before for agitating, spent an hour in jail and was later released on a fifteen-dollar bond.[38]

Wobbly soapboxers faced attacks in other communities as well. On August 17 at Drumright, hecklers jostled organizers Charles Clinton and A. A. Rice while young boys pelted them with rotten eggs. When the men spoke the next day, eggs again flew, this time striking town residents. A small riot ensued, and police told Rice not to speak in the town again. At Muskogee in October, Brink and four sympathetic Socialist Party members—including Frank Ryan, who served as secretary of the Tulsa OWIU local in 1917—engaged in a free-speech fight and were repeatedly arrested. A local judge ruled the men had a right to speak even if they drew crowds that blocked the street, but police again arrested them when they tried to assert their rights. "You all know what this means," Clinton wrote *Solidarity*. "If they can stop us from speaking in the streets, they can stop us from speaking in the hall; if we can't speak, we can't organize."[39]

Ironically, while the attempts to suppress street-speaking failed to stop the organizing campaign, an unrelated moratorium on drilling succeeded. Many small producers and independent oil companies saw prices for their crude drop in 1914, while pipeline rates remained high. Rather than pay, the small producers worked out a "gentlemen's agreement" to restrict production, the state attorney general agreed not to enforce antitrust laws, and in July the corporation commission banned the drilling of new wells. By autumn 10 million barrels of crude were in storage at Cushing. The price had dropped to forty cents a barrel as producers had to sell the oil at below market rates before it deteriorated.[40]

With this market collapse, the need for workers declined, just as it had in 1907. The pipeline companies stopped laying lines, and many pipeliners gradually drifted from the fields. The sudden loss of members forced the Tulsa local to disband, while the Drumright branch barely managed to hang on.[41]

Despite the losses, the OWIU continued activity in the oil fields in 1915, and delegates expected a mild revival of drilling activity. Delegate Walter Smith noted the Cushing and Drumright areas suffered from labor shortages, especially of tank builders, which the IWW could exploit. An AFL tank builders' strike had been mildly successful as a result, and even a 300-member independent union of the normally elitist drillers and tool-dressers had formed in Cushing by April. Smith believed the IWW could make gains if it made a systematic effort.[42]

Encouraged by Smith's reports, the General Executive Board chartered a new local at Cushing on June 21, 1915, and agitators went out into the few remaining pipeline camps. They achieved little success. In July, an IWW-led strike at a Prairie Pipeline camp failed when Prairie officials elected simply to discontinue work rather than meet a 25¢-a-day pay increase demand.[43]

Like their harvest counterparts, Wobblies also found towns and cities increasingly hostile. In late November or early December Oklahoma City police arrested eleven IWW members en route to the oil fields and charged them with vagrancy. In a cursory trial, a municipal judge refused to let the men testify in their own defense and found the men guilty, fining them $99 each and sentencing them to ninety days on the county road crew. The men gained their freedom only after making a habeas corpus request to the Criminal Court of Appeals. Appeals Justice Thomas H. Doyle agreed that the men had been denied due process by not having counsel or the right to present a defense, and he ordered their release. Criticizing the municipal judge for exceeding his authority, Doyle wrote, "The court shall not arbitrarily take citizens' liberty from them under the provisions of city charters in violation of state laws and the fundamental laws of mankind."

The verdict was certainly a victory for the IWW, and even the *Oklahoma Labor Unit*, a pro-AFL newspaper, said Justice Doyle had a real understanding of liberty and justice. But Oklahoma City officials said they would continue to arrest men on both vagrancy and trespassing charges. One IWW member wryly commented that it seemed to be a scheme to supply men for the county road crew.[44]

Despite the legal victory, the 1915 campaign was a failure. The IWW had clearly overestimated the extent of activity in the oil fields and underestimated both the willingness of oil companies to wait out work stoppages and the lengths to which municipal authorities would go to arrest IWW members. All of that would change in 1916 as two factors boosted oil production. The first was oilman Harry Sinclair's defeat of Standard Oil in a price war. Sinclair held oil in storage until a shortage tripled prices, and he plowed the profits into a pipeline and refinery system to compete with Standard. Second, the European demand for oil to fight the First World War forced up prices. While both factors increased activity and jobs, the second meant the federal government would do whatever was necessary to protect oil production once America entered the war. That boded ill for the IWW.[45]

The success of the 1916 AWO harvest campaign also renewed the IWW's interest in the oil fields. Both *Solidarity* and the *Industrial Worker* ran articles comparing the huge profits of oil companies to the low pay of the workers and the high cost of food and housing in the Oklahoma oil towns. "While you are inhaling poisonous gases, up to your shoe tips in filthy muck, the bosses are wallowing in luxury, eating the choicest of foods and drinking champagne," one delegate wrote.[47]

By fall, AWO delegates were operating from the existing local in Cushing and from new ones in Tulsa and Drumright. The Drumright IWW local was especially interesting as many members held both AFL blue cards and IWW red cards, using the latter essentially as "free" railroad passes. Enough oil workers responded to the union that the AWO opened other locals in Healdton and New Wilson (renamed Wilson in 1920) in Oklahoma, and in Augusta, Kansas. On January 1, 1917, the GEB officially chartered the new branches as Oil Workers Industrial Union no. 450. "The opportunity for putting a dent into John D.'s oil can is good," one delegate said.[48]

The new Tulsa local, formed in November 1916, picked up where the old branch left off and began street agitation under the direction of delegates Ted Fraser and James Duffy. Organizers called for a five-dollar daily wage, double pay for overtime or dangerous work, the eight-hour day, and no discrimination against IWW members. Saying they had nothing to hide, branch officials invited the public to visit the various halls and learn about the IWW.[49]

Few, if any, Tulsa officials or oilmen took up the invitation. To stop street-speakers, the city commissioners approved a new ordinance mandating a $100 fine for loitering, while the *Oil and Gas Journal*, a Tulsa-based petroleum industry publication, claimed the IWW was openly advocating class war. Drumright authorities also worried about the reappearance of the Wobblies. In early February 1917, police there arrested Arthur Boose, a longtime AWO job delegate, for vagrancy, although he proved in court that he was a salaried organizer for the IWW.

Known as the "Old War Horse," Boose was a veteran organizer who worked among the Finnish iron miners in the Mesabi Range of northern Minnesota. Facing a trumped-up murder charge there, Boose fled and, under the name Arthur Fritz, was driving a team on railroad construction jobs when the General Executive Board sent him to Oklahoma. Despite the threat from authorities, Boose remained irascible. At Drumright he launched into a tirade against the police and the judge, the result of which was a $25.50 fine for contempt. About fifty union members attended the trial, which may have discouraged police from making further arrests.[49]

The spring witnessed increased union activity by both the IWW and the AFL, the latter mostly in Texas and Louisiana. Union activity grew particularly after the oil companies began demanding concessions from labor at a time when the firms were charging the government $3.00 a barrel for oil that cost about 40¢ per barrel to produce. Pipeliners, many of them Wobblies, struck for forty-five days and won a $4.00 daily wage and the eight-hour day from a Shell subsidiary. A Wobbly-led strike at a Prairie Pipeline camp in Augusta, Kansas, gained paid transportation back to town at the end of the work day and an hour for lunch, though the workers complained that the company food was still poor. These successes encouraged union membership in other regions. In Goose Creek, Texas, some oil workers joined the IWW, although that state was predominantly AFL territory. In Kansas, an influx of Mexican and Eastern European workers in the fields led to a call for delegates who could speak Spanish and Bulgarian to help recruit those men. By autumn, the OWIU could reasonably claim it had raised pipeliners' wages to $3.50 a day and increased roustabouts and common laborers' salaries to $100.00 a month.[50]

But events both outside and within the IWW plagued the organizing campaign. In March 1917 Haywood and the GEB, long jealous of the AWO's power, stripped the organization of its nonagricultural members. Although the oil workers technically remained under the AWO's control, the other branches were placed in separate industrial unions controlled by the Chicago headquarters. The AWO itself was renamed Agricultural Workers Industrial Union (AWIU) no. 400. A month later the United States declared war on Germany and began preparations that led to the prosecution of the IWW leadership and the suppression of the OWIU.[51]

With the GEB firmly in control of the IWW, Haywood began measures to place his supporters in the oil fields. One of his actions was to appoint Phineas Eastman OWIU branch secretary first in Drumright and then in Augusta, Kansas. Slightly built, with graying hair and gray eyes, the forty-six-year-old Eastman was a Mississippi native and a ten-year member of the IWW. He had worked a variety of odd jobs before joining the Brotherhood of Timber Workers and taking part in the Merryville strike. Like other BTW members, he drifted into oil work after the strike failed; but unlike other delegates, who preferred to play "safety first" and organize quietly, the contentious Eastman openly advocated sabotage. In May 1917, he allegedly wrote a letter in disguised handwriting warning the Augusta Chamber of Commerce that any attempt to suppress the IWW would be met with dynamite and nitroglycerine.[52]

The letter reached federal authorities, who feared widespread resistance to the newly approved Conscription Act. Although the IWW opposed all wars, it never took a formal stand on either the war or the draft, perhaps because the leadership realized the consequences of such an action. But that did not stop federal officials from undertaking surveillance operations and plotting raids on IWW offices in preparation for a full-scale prosecution of the union. In addition, oil companies hired additional private police and labor spies to infiltrate the OWIU. The result would be a war, but one the IWW was ill-prepared to fight.[53]

Overall, the IWW's campaign to organize oil workers stands as one of the most important, and dangerous, decisions the union ever made. The emerging petroleum industry would soon be the backbone of a new wave of industrial development in the United States, a fact the Wobblies certainly understood. By recruiting oil workers, the IWW moved into a major new area, one that was totally controlled by corporate forces openly hostile to labor and highly exploitative of it. The campaign also eventually help created splits within the union between those favoring centralized control and those seeking a more decentralized structure with autonomous branches. The AWO's campaign had paradoxically demonstrated the need for decentralized organization while increasing the centralized control of the GEB. The oil field campaign drew on the AWO strategy and would, in the long run, only heighten the differences.

But the petroleum industry campaign was a product of the AWO's success and demonstrates the growing confidence of the union as membership and financial resources grew. It reveals the increasing commitment of the IWW to recruiting migratories. Yet, while the union could control the harvest labor supply to some extent, organizing oil workers posed other problems. The immediate need for farm hands by single farmers or locally-based farming companies gave the AWO some leverage in demanding better hours and pay. But the need for oil workers depended on factors outside of the Wobblies' control. National demand for oil affected the need for pipeline construction, while state regulations could determine the amount of petroleum drilling. In addition, the nature of the oil workers' culture made organization difficult. While the harvest hands operated as a "level" workforce, the oil workers developed a hierarchical one in which some jobs held special status, thus creating animosities between the workers that employers could exploit. The IWW enjoyed success with pipeliners probably because they were more "level" than other oil workers and because they were isolated in camps like those in the lumber industry.

More important, though, the campaign would prove dangerous as it exposed the IWW to more powerful forces than the union had previously encountered. The oil towns and drilling operations were almost totally controlled by powerful corporations that drew on large resources, both financial and otherwise. In addition, the growing importance of petroleum to the war effort placed the power of the state and federal governments squarely behind the oil companies.

The oil field presence of the IWW and its violent reputation were both a cause for alarm and an opportunity to destroy the labor movement in the oil industry. In Oklahoma, with its strong radical movements, the businessmen's fears were undoubtedly greatest. The chance to deal with the IWW and radicalism in general came in the wake of the August 1917 Green Corn Rebellion, an ill-timed and ill-conceived tenant farmers' uprising. The authorities and the press quickly pinned blame on the IWW, whose name had come to represent all those who opposed the war or questioned the supremacy of the American economic system. In the wake of the rebellion, the *Tulsa Democrat* declared that the "Industrial Disturbers of the World" would pay for their opposition to the war. "They do not reckon with the strong arm of the government," the paper declared. Indeed, the strong arm was not long in coming.[54]

4

With Folded Arms? Or with Squirrel Guns? The IWW and the Green Corn Rebellion

REGARDLESS of the IWW presence, by the summer of 1917 rumors of a radical uprising had spread throughout Oklahoma. Resistance to the new federal conscription law had increased, and law enforcement officials and newly formed councils of defense feared violence, especially from the Working Class Union (WCU), a radical tenant farmers' organization. Events came to a head on August 2 when the Seminole County sheriff and his deputy were ambushed near the Little River. Within hours raiding parties had cut telegraph and telephone lines, burned railroad bridges, and allegedly dynamited oil pipelines near Healdton. The next day members of the WCU gathered on a Pontotoc County farm in preparation for a march on Washington that would force President Woodrow Wilson to end the draft. But the revolt proved short-lived after a hastily formed posse attacked the farm. When all was said and done, 3 men were dead and almost 450 others were arrested for their participation in the rebellion. About 150 were convicted, and many served lengthy prison terms at the state prison in McAlester and at the federal penitentiary in Leavenworth, Kansas.[1]

The revolt fueled antiradical sentiment within Oklahoma, with the first casualty being the state's reform-minded Socialist Party. The Socialists' electoral strength lay with the debt-ridden tenants, but the party had disavowed both violence and the WCU. Still, the fact that some WCU members were Socialists only fueled the reactionary backlash. The biggest attacks, however, were leveled at the IWW.

Like many Americans, Oklahomans had heard tales of IWW members burning wheat crops or driving iron spikes into trees to ruin lumber mills. Other rumors alleged that Germany had paid the IWW to disrupt the American economy. Claiming the WCU was simply the IWW by another name, Oklahoma newspapers began a jingoistic campaign against the Wobblies. Some editors openly called for lynch mobs to deal with those whom one United States senator styled "Imperial Wilhelm's Warriors."[2]

But in reality the IWW took no part in the Green Corn Rebellion and had only a slender, indirect connection to the Working Class Union. In fact, the WCU was a product of the Wobblies' refusal to organize tenant farmers and grew out of the farmers' misinterpretation of IWW ideology. Much of this misunderstanding arose because of the increased harassment and intimidation of Oklahoma Socialists. As a result, radical farmers moved toward direct action politics to push land-tax reform and resist moves to disfranchise tenants. Many militants—particularly those influenced by the more radical Texas SP—found the Wobblies' syndicalism appealing, especially the evangelical, millenarian style of its rhetoric. Such an ideology spoke to both their frustrations and their own traditions of self-preservation and self-determination.

The spread of IWW ideas among the tenants suggests that the Wobblies possessed more and broader-based support and influence than mere membership roles would indicate. But the IWW rejected the tenants, in part because admitting the farmers would conflict with the union's own ideology. Seeing tenant farmers not as workers but as capitalists who hired labor, many IWW editorialists and ideologists viewed them as holding to outmoded ideas, especially as to private property, and urged the union to refuse any alliance with employers, no matter how feeble. Yet the IWW failed to see that the tenants and small farmers had already become wageworkers in all but name, and wageworkers sympathetic to Wobbly goals.

In response to the rejection, some Oklahoma tenant farmers turned to the WCU, blending a muddled industrial unionism with traditional southern forms of countervigilantism, self-defense, and opposition to conscription. The IWW, as a result, failed to seize an opportunity to broaden its appeal, but instead received the blame for anything the WCU did. Indeed, the WCU and the Green Corn Rebellion gave federal authorities the legal means to procede against the IWW nationally.

The rise of the WCU and other agrarian syndicalisms came in part from the collapse by the end of the nineteenth century of the "agricultural ladder" myth. Under this long-accepted concept, a farmer began as a hired hand, then saved enough money first to rent a farm and then to buy one of his own. But as the twentieth century dawned, more and more farmers, rather than rising to farm ownership, found themselves pushed down into the ranks of the tenant or even the migratory agricultural laborer. The causes of this decline included the exhaustion of the public domain, an increase in both legitimate and fraudulent land speculation, and a growing trend toward corporate ownership of agriculture.[3]

The effects proved most severe in Oklahoma. Once seen as the farmer's last, best hope, it all too quickly became the realm of landlord and tenant, with the landlords as beneficiaries. Operating behind a façade of legal formalities, many had come to control Indian allotments and monopolized the best farmland. Generally southerners and business progressives, they quickly consolidated holdings into large-scale farms for mass cotton production. Tenants were the preferred form of labor, partly due to the nature of cotton cultivation and partly out of fears of an actual wage-earning labor force in the countryside.[4]

Speculation and outright fraud drove Oklahoma land values up by 246 percent between 1900 and 1910, mostly because land held for speculation proved more valuable than the crops grown on it. With land so expensive, farmers could not afford a farm of their own and were reduced to tenantry. While less than .7 percent of the region's farmers were tenants in 1890, by 1915 more than 50 percent—104,000 of the state's 200,000 farmers—

were tenants. Of the land owners, 80 percent were heavily mortgaged.⁵

Most tenants were young: of farmers age 24 or younger, 76 percent were tenants, while 45 percent between the ages of 25 and 33 rented their land. Unlike most tenants in the South, the bulk were white. Few would ever own land, as merchant credit and the crop lien system dissipated whatever capital a tenant could scratch together. Less and less an independent economic actor, the tenant became an agricultural laborer whose earnings amounted to a subsistence wage. Because the Oklahoma soil proved poor even for subsistence growing, tenants had to spend from fourteen to eighteen work hours per acre—twice the rate of Mississippi or Louisiana sharecroppers—to compensate for lower yields. Children as young as three were expected to pick cotton, but even with the entire family working, tenant farmers often took other work—including joining the wheat harvest or doing the most menial oil field tasks—in order to make ends meet. Compounding their problems, the tenants' diet—usually salt pork, corn meal, and molasses—and unsanitary living conditions left them susceptible to maladies such as pellagra, typhoid, hookworm, and malaria.⁶

Exploited by the landlord and the merchant, many tenants turned to the Oklahoma Socialist Party to improve their lot. The Socialists offered political organization, a plan to expand the public domain for tenants' use, a graduated land tax to reduce cut-throat speculation, and a cooperative marketing proposal. But other tenants rejected the ballot box route to revolution and moved toward protosyndicalism or agrarian syndicalism. The drift increased as more and more landlords refused to rent to Socialist tenants. The Oklahoma Renters Union and the Farmers Emancipation League, as well as the Texas-based Land League and the Farmers' and Laborers' Protective Association (FLPA), defined themselves as revolutionary, direct-action organizations and were precursors of the WCU.⁷

The SP's left wing generally organized these groups, but some tenants, especially after 1914, had contact with the IWW either by working the wheat harvest or by working in the oil fields. Others heard Wobblies such as Frank Little speak to their Socialist locals. The IWW's ideology probably appealed to the tenants for several reasons. Certainly the Wobblies' revivalist style, with its apocalyptic imagery of the general strike, coincided with the Pentecostalist Christianity of many tenants, while the IWW's use of sabotage—in heavily distorted fashion—fit well with the southern tradition of self-defense and active resistance. But the Oklahoma tenants were harldy "primitive rebels" opposing industrialization. Many had become participants in the process itself as they joined the wheat harvest or drifted to the oil fields and underwent experiences similar to those of the rural folk who had formed the first wave of an industrial working class. In IWW members, the tenants found men who shared those experiences, unlike the often middle-class Socialist organizers. These encounters may have heightened the tenants' feelings of separation from society as a whole. Yet if the traditional ties of the tenants were to family or clan, the exposure to both Socialist and IWW ideas gave the tenants a sense that they were part of a larger "clan" of the dispossessed. Significantly, an armed offshoot of the IWW-affiliated Brotherhood of Timber Workers in Louisiana called itself the Clan of Toil and was composed of both those tenants who had both been forced into lumber jobs and those who had remained on the land.[8]

On the surface, then, the tenants and small farmers and the IWW should have been drawn together. The farmers saw themselves and the workers both as producers victimized by the nonproductive money barons and captains of industry. But while many individual Wobblies sympathized with impoverished tenants, the IWW refused to recruit farmers of any sort because they were not wageworkers and because they represented the last vestiges of feudalism in America. Much of the Wobblies' opposition derived from their experiences with the Knights of Labor, in which the rural and middle-class reformist elements repeatedly clashed with the labor union assemblies.[9]

Early in its history, IWW members such as Ethel Carpenter of Oklahoma had described farmers as capitalists at heart, no different from other employers. One IWW writer noted that the farmer's views on property and exploitation differed little from those of other businessmen, and the farmer was only angry that he was not getting his fair share. William Mead, in a *Solidarity* article called "Keep Out the Farmer," even compared the farmers to lawyers, a group excluded from the Knights. "For as the lawyer, the farmer is a parasite on the worker's back, and we should apply to all parasites the same terms, and advise them to take a back seat for the good and welfare of our movement," Mead declared. For most of these writers, the problem would be resolved when capitalism consolidated farms into huge enterprises and converted the small farmers into wage earners.[10]

The tenant question came to a head in the wake of the 1912 and 1913 strikes of the BTW. BTW leaders, such as Covington Hall, saw the recruitment of sharecroppers as essential in the union's violent dispute with the lumber trust led by the Texas-based John H. Kirby Lumber Company. Hall, who edited *The Lumberjack*, the BTW newspaper, blamed the failure on the IWW's refusal to understand the southern labor situation and its rejection of the tenant farmers, many of whom were related to the "redbones," or poor whites, who constituted the bulk of the timber workers. In addition, Hall noted that both black and white farmers and merchants who held populist values supported the BTW. In several articles reprinted in both *Solidarity* and the *Industrial Worker*, Hall called for the active recruitment of these groups because they were as much rebels as were the workers.[11]

But the IWW refused to budge. In response to Hall, the editor of the *Industrial Worker* again argued that tenant farmers were employers who relied on migrant labor. If allowed in as members, the tenants would create class antagonisms. Further, the writer noted, if such antagonisms did not occur, it would blur the meaning of the class struggle and destroy the purpose of the IWW. Another IWW writer, Ernest Griffeath, accepted Hall's argument that tenants could be rebels, but he added that small traders had also once been rebels: "[B]ut they fought for the freedom of the budding capitalist class; they fought so they might exploit the wage workers in place of the feudal lords." Griffeath did not believe giving tenants union cards would make them revolutionaries, nor would it prevent real revolutionaries among the croppers from studying the principles of industrial unionism. Other Wobbly writers argued that small farmers opposed centralized, large-scale industry and thus were in conflict with the industrial workers' interests.[12]

Yet some members of the IWW saw a potentially fruitful alliance with tenants. Both E. F. Doree, later an important figure in the AWO, and T. F. G. Dougherty urged the creation of an auxiliary organization for farmers. B. E. Nilsson agreed that tenants should not be allowed in the union, in part because they could not use strikes or sabotage to protect their interests. But he noted that many tenants also worked wage jobs in the lumber, turpentine, and railroad industries. Because the IWW was not just a union but rather part of the coming revolution, it should encourage tenant unions and maintain friendly relations with them, Nilsson contended. The workers after all had a lot more in common with the tenants when organized in a union than when organized in a political party. Like Doree and Dougherty, Nilsson argued that it was wrong to see tenants as capitalists. "[T]o anyone who is familiar with their hard struggles for a bare existence, this classification seems very absurd. Whatever reason there may be for barring them out of the union, it will not be on account of the capital they possess."[13]

But Nilsson's position was in the minority. For the IWW, the issue was settled: no tenants. That left some individual Wobblies such as Hall frustrated. "[T]hose sodbusting and sharecropping farmers were workingmen, and, more important, were the fathers, uncles, mothers, aunts, brothers, sisters, and cousins of the men working in the oilfields, forests, mills, and mines of the South," Hall later wrote.[14]

Faced with the IWW's lack of interest, Hall and several BTW members formed the Working Class Union in 1913. The organization's charter was written by "Uncle" Fred Freeman, a seventy-year-old Mississippi populist and veteran of the Knights of Labor. The charter first appeared in Hall's magazine *Rebellion*, published in New Orleans, and was reprinted on WCU membership cards. It declared the signers had pledged to be loyal to their class in the struggle between workers and capitalists and to work toward the one big union, just like the IWW. "We will not scab or knock on any other working man or woman. We will do all in our power to help every worker to win his demands for a better living," the charter stated, adding that members might serve in any capacity to bring about industrial solidarity.[15]

Compared to the IWW's declaration that "[t]he working class and the employing class have nothing in common," and that the two would be engaged in struggle until the workers took control of society, the Working Class Union charter seems rather tame. But while the IWW constitution spoke of revolutionary struggle, the union rejected violence. The WCU charter, couched in less threatening terms, did not reflect that organization's far from passive actions. As Hall wrote, the WCU did not believe in the IWW's "mighty power of folded arms," but both accepted the dictum "A scab has no rights an honest working man is bound to respect."[16]

The WCU initially tried to secure help in Louisiana, but, failing that, sent a BTW member named John E. Wiggins to southeast Oklahoma and western Arkansas. In 1914, the WCU, now under the leadership of Dr. Wells LeFevre, set up a headquarters at "Hobo Hollow," near Van Buren, Arkansas. Many have erroneously assumed that this constituted the actual creation of the union. Under the direction of LeFevre, an Arkansas Socialist and friend of Father Thomas Hagerty, the WCU spread to Kansas, northern Texas, and even Nebraska, but it found the most fertile soil in southeastern Oklahoma and along the South Canadian River, with locals formed in Hughes, Seminole, and Pontotoc Counties. Its leaders would eventually claim a membership of 35,000 in Oklahoma alone, although this is probably an exaggeration.[17]

The collapse of cotton prices as the war in Europe closed off those markets aided the WCU's efforts. The collapse, from nearly twelve cents a pound to just under seven cents, came at a time when the local price of cotton was only half the estimated production cost. But while crop prices and land values dropped, loans and mortgages did not. Foreclosures increased, and landlords demanded a greater share of the crop to cover their own losses.[18]

The WCU quickly became the vehicle for the tenants' frustrations. Initially operating like the Socialists' Oklahoma Renters' Union of 1909, the WCU's goals differed little from the SP's own. Formally approved at a January 1916 WCU convention at Sallisaw, they included the abolition of rent and the wage system; the eight-hour day; workers' compensation and old age pensions; child labor laws; free school textbooks; and the initiative, referendum, and recall in the state government. Unlike the IWW, the WCU did not reject political action and helped elect an attorney named L. C. McNabb to a Sequoyah County judgeship. Members also came to McNabb's aid when local authorities accused him of embezzlement while in office and tried to have him disbarred.[19]

But as cotton prices continued to fall and an overall agricultural depression set in, the WCU shifted toward more violent forms of direct action and social banditry. The trigger was an outbreak of Texas fever among the state's cattle in 1915. State agriculture officials ordered a dipping program to eradicate the ticks believed to carry the disease. Those farmers who refused to cooperate faced seizure of their livestock by the county sheriff. Small farmers and tenants accused authorities of helping the large herd-owners, whose cattle were actually spreading the disease, and they also claimed the dip was improperly mixed and killed the cattle. Because the loss of any livestock spelled economic disaster, the WCU struck back when authorities enforced the program. Beginning in September 1915, night riders dynamited dipping tanks in Pontotoc, Sequoyah, and Muskogee Counties; burned the barns of those county commissioners who approved the program; and whipped a tenant who had signed a contract with a landlord blacklisted by the WCU. The night-riding activity came to be known within the WCU as "sending the Jones Boys," which later led the press to identify one WCU local as the "Jones Family" and to assume incorrectly that they were a separate organization, an error perpetuated by several historians.[20]

The night-riding campaign led many contemporary observers to contend that the WCU was tied to the IWW. But night riding owed far more to traditional agrarian resistance to outside control than it did to the Wobbly concept of direct action. With the exception of a farmhands' strike at Moffett in May 1916, the WCU never used the IWW's more sophisticated brand of sabotage. A form of social banditry, night riding had long been common in both Texas and Oklahoma. As early as 1908 in Oklahoma, night riders were active near Idabel, and in 1910 authorities arrested five farmers, probably tied to the Renters' Union, for night riding. The WCU also drew on other traditional symbols of southern self-defense, including draft resistance. In fact, after Congress approved conscription in 1917, the WCU adopted actions similar to those of the Arkansas farmers who had opposed the Confederate draft during the Civil War, including signs and countersigns and an oath sworn with one hand on both a pistol and the Bible. It even used the old Civil War southern draft resister's slogan "Rich man's war, poor man's fight."[21]

But the WCU's direct action, traditional as it was, still fed the anxieties of those citizens who feared radicalism in any form. To their minds, the violence of the WCU was precisely that of the IWW. Victor Harlow, editor of *Harlow's Weekly*, summed up those feelings: "Socialism, I.W.W., W.C.U.—these are flaming warning signals which no wise statesman will ignore and which no shrewd politician can afford to ignore." While Harlow meant the phrase as a call for land reform, most readers undoubtedly took the statement at face value. But public worries declined somewhat in 1916 after cotton prices rose slightly and the state legislature passed an anti-usury law.[22]

The lull lasted until the United States entered the First World War in April 1917 and Congress passed the Draft (or Conscription) Act. Poor southern whites had opposed conscription since the Civil War, usually because farmers' sons were infrequently exempted and drafted hired help was rarely replaceable. Oklahoma's small farmers viewed the draft as an added burden. Influenced by the Socialists, they quickly saw the conflict as a bosses' war to protect the interests of bankers such as J. P. Morgan.[23]

The WCU only needed someone to articulate their anger, and they found him in an agitator named Henry H. "Rube" Munson. Munson has long been considered one of the direct links between the IWW and the WCU. Writers both sympathetic to and hostile to the WCU have accepted the fiction that the IWW leadership sent Munson, who had fled a criminal indictment in Chicago. This is simply not true. Munson was a former lead and zinc miner from the tristate town of Seneca, Missouri, where his family lived with his wife's parents. He had organized for the WCU for two years by 1917. It is highly unlikely that Munson was a Wobbly as the IWW had virtually no presence in the Tri-State region until the early 1920s. He was indeed indicted in Chicago, but along with more than 160 others—many of them non-Wobblies—named in the massive federal indictment of late 1917 that accused the IWW of hindering the draft and violating the Espionage Act. Four other Oklahomans—two of them active Wobblies—were also indicted, but only one, Walter M. Reeder, had any contact with the WCU, and indirect at that. In addition, that indictment occurred after the Green Corn Rebellion, not before it.[24]

Federal authorities even conceded Munson was not a Wobbly. Redmond S. Cole, then an assistant U.S. attorney for the western district of Oklahoma, told the head of the federal board of parole that he believed Munson was "an I.W.W. at heart" and had alone conceived of leading the WCU to active draft resistance. Additionally, state newspapers never said Munson belonged to the IWW. The *Daily Oklahoman*, for example, only compared his direct actionist stand to that of the Wobblies. Finally, the Federated Council of the Churches of Christ's 1920 report on conditions at the federal prison in Leavenworth, Kansas (where the WCU leaders, including Munson, were held), noted that none of the WCU's members belonged to the IWW.[25]

Whatever the case, once the draft was approved, Munson called on WCU members to resist conscription, and he apparently believed he could obtain the IWW's support. One former WCU member later testified that Munson had said forty-eight organizations, including the IWW, would join to end the war. Another WCU member, Homer C. Spence, made a similarly fanciful claim. Spence, who served as Munson's lieutenant, said two million Wobblies would join with the WCU and the Farmers' and Laborers' Protective Association of Texas to capture "Kaiser Wilson." He asserted that members in Germany would capture Kaiser Wilhelm at the same time and end the war.[26]

All of this was news to the IWW, especially to the Agricultural Workers Industrial Union (AWIU). At its 1917 spring convention held in May in Kansas City, the AWIU pointedly refused to affiliate with the WCU. The leadership argued that the Working Class Union was not a union because, besides tenants, it also included doctors, lawyers, and merchants. In addition, the IWW had grown worried about assaults on its own members, rabid criticism from the nation's press, and growing federal surveillance. Trying to avoid taking any position on conscription, the union clearly feared allying itself with an organization such as the WCU, which was already known for using violence.[27]

But the refusal to support the WCU failed to aid the Wobblies. Oklahoma newspapers increasingly used IWW and

WCU in the same breath and alleged they were part of a nationwide conspiracy. Government spies and paid informers had already penetrated both organizations, and authorities soon viewed the WCU as a means to destroy the Wobblies. They saw the link not in Munson, but in Walter Reeder, an IWW member tied to the WCU and a third organization, the ill-conceived Universal Union (UU).²⁸

A Socialist and a Wobbly, Walter Reeder had worked for almost ten years as an oil driller. Married, with two children, and a longtime resident of New Wilson, Reeder was hardly the stereotypical Wobbly. By 1917, he had become an OWIU organizer in the Healdton Pool near Ardmore. It was through this position and his membership in the Socialist Party that Reeder may have had contact with individuals tied to the WCU, although he himself never joined. Why Reeder decided to help form the Universal Union remains a mystery, although it seems reasonable to assume he opposed the AWIU's rejection of the WCU.²⁹

Whatever the reason, Reeder met with at least six other men, including veteran radical SP member J. T. "Tad" Cumbie, on July 17 and 24, 1917, at a farm near El Reno. The men allegedly held a secret meeting of the UU and passed a resolution—which they supposedly burned after approving it—opposing the draft and selecting delegates to contact the IWW, the WCU, and what they called the "Revos"—radical elements in the Pacific Northwest—to plan the overthrow of the federal government. They adopted a military-style structure and named Reeder the state general. Cumbie apparently went to Colorado to parlay with the "Revos," while Reeder went to Chicago, where the IWW leadership must have rebuffed him. On his return, however, Reeder—like the WCU's Homer Spence—claimed that two million Wobblies would help them capture the state and local governments, a wild assertion considering that the IWW had nowhere near that many members.³⁰

An informer exposed the UU's fanciful plans, and federal agents arrested the entire UU membership in the summer of

1917 on charges of conspiring to obstruct the draft. Reeder, like Munson, was initially included in the mass Chicago indictment. Those charges were dropped, but Reeder and the other UU members were reindicted in 1918 and convicted in federal court in Enid that fall. Although the Universal Union was clearly an inconsequential organization, it offered federal prosecutors something more. The trials of the UU members, eight WCU members from Pottawatomie and Cleveland Counties referred to as the "Jones Family," and the leadership of the Texas FLPA became test cases for the federal government. By gaining convictions on fairly flimsy charges, prosecutors felt convinced they could proceed against the IWW. The tragicomic Green Corn Rebellion only aided the prosecutors' cause.[31]

Tensions increased in June and July of 1917 after the press and state authorities accused the WCU of causing a gas explosion at Kusa and of dynamiting a water storage tank at Dewar and sewer mains in Henryetta. The arrest of draft resisters—not all of them IWW or WCU members—continued unabated, and by July, Munson and Spence and the "Jones Family" were in federal custody. WCU charter member John Wiggins was also arrested, but in Washington State, where he was working as a waiter at an army camp, far from the events he had helped set in motion in 1914.[32]

Then, on July 27, 1917, an aging farmer named John Spears raised a handmade Socialist flag and an American flag over his property near Sasakwa along the South Canadian River. There gathered other men opposed to draft, prepared to lived on barbecued beef and roasted green corn on their march to Washington to confront "Big Slick"—President Woodrow Wilson. On August 3, the day after ambushing the sheriff and his deputy, the rebels launched their plan, only to confront a hastily assembled posse. A few shots were fired, but the great rebellion ended bare hours after it began, with 3 men dead and almost 450 prisoners. Of the would-be rebels, 266 were not indicted and were released. Another 184 faced charges, and approximately 150 were convicted or pleaded guilty, with about half receiving jail terms ranging from sixty days to ten years. The leaders, such as Munson and Spence, received the heaviest sentences. There were some protests over the leniency of most of the sentences, but the Department of Justice had earlier informed W. P. McGinnis, the U.S. attorney for Oklahoma's eastern district, not to dignify the uprising as treason. The authorities viewed the rebels as ignorant small fry, and they sought bigger fish, namely the IWW and its General Executive Board.[33]

The government's intentions became clear during the trial of the so-called "Jones Family," held in Enid in September 1917. These men, WCU members from Cleveland and Pottawatomie Counties, were accused of conspiring to obstruct the draft, and John Fain, the federal prosecutor, repeatedly pressed the witnesses to connect their alleged conspiracy to the IWW. When the defense attorneys, Patrick S. Nagle and John J. Carney, objected, Judge John H. Cotteral, a vocal critic of draft resisters, surprisingly sustained the objection, noting that the prosecution had failed to provide any evidence to link the two organizations. In addition, three defendants—Clure Isenhour, his brother Obe, and John Shirey—not only denied they were Wobblies but added they did not even know what the IWW was or what it stood for. The federal government later made one more unsuccessful effort to link the IWW and WCU through Walter Reeder during the Wichita trials of several OWIU members in 1918.[34]

But the need to show a direct connection was unnecessary. The trials of the "Jones Family," the Green Corn rebels, and the Universal Union proved their worth to the federal authorities by producing convictions for conspiracy under the Draft Act and Espionage Act. The success of those trials clearly gave the justice department the necessary confidence and know-how to pursue indictments against the IWW in Chicago, Wichita, and Omaha, Nebraska. The WCU, therefore, inadvertently provided the authorities with the hammer they could use to smash the Wobblies, and that may have been the most important connection of all.[35]

That the IWW's refusal to recruit tenant farmers would come back to haunt them is the greatest irony of the Green Corn Rebellion. The union's insistence on recruiting only wageworkers meant it failed to see the tenants as the agricultural proletariat they were quickly becoming. But the Wobblies' rejection did not mean the tenants were unreceptive to IWW ideas. As some IWW members realized, many tenants and small farmers took wage-paying jobs in industry and suffered the same hardships as the hoboes who composed much of the IWW. Both wageworkers and tenant farmers were at the mercy of modern business practices that prevented them from gaining even the smallest margin of economic self-sufficiency. Ostracized from a society that viewed them as ignorant, lazy, and even biologically inferior, they clearly sought a saviour that could show them how to rise up against their oppressors and bring about a new social order. The hoboes looked to the IWW; for the Oklahoma tenants, their two possible means of salvation—the Socialists and the IWW—proved inadequate. Instead they were forced to create their own organization, the Working Class Union, which fed their hostility toward the capitalist system. When that system began to create further burdens, such as the dipping campaign and the draft, they turned to older forms of resistance that drew on traditional values such as the need for self-preservation, the right of revolution, the sovereignty of the people, and the need for vigilance. To this they added more sophisticated ideas drawn from the Socialists and the Wobblies.[36]

But for a populace, both in Oklahoma and nationwide, in the midst of a wartime patriotic frenzy and awash in a steady stream of news stories about the "lawless" IWW, the finer distinctions among the Socialists, the Wobblies, and the WCU were meaningless. Already hostile to organizations such as the IWW, the authorities irrationally saw the radicals as part of a widespread conspiracy funded by the Germans. In such an emotional atmosphere, the actual crimes of the WCU were seen as only a hint of the larger threat it posed, and the IWW became the perfect scapegoat.

5

War and Repression

THE First World War proved both a boon and a bane to the IWW nationally and in Oklahoma. The war gave the farmer prosperity, thereby strengthening the AWIU, which could threaten the farmer's prized income with work stoppages. In other industries, the IWW—like many AFL unions—benefited from employers' fears of rising costs and labor shortages, which temporarily reduced employer resistance to unionization. Between 1916 and 1917, the IWW increased its membership from 40,000 to 100,000 members.[1]

But most of those benefits accrued before American's entry into the conflict. Once war was declared, the resulting war effort proved a triumph for the incorporation process. The need for vital materials meant government and business formed a closer alliance than ever before and made more concrete the bureaucratic and business-oriented modernization of the state and industry that was the hallmark of Progressivism. The results were twofold. First, business and industry gained the official imprimatur of the government and could call on its support to guarantee production. Second, the government could rely on business to help mobilize the public behind what had generally been an unpopular war. The government, through such organizations as the Committee on Public Information, began to shape the limits of permissible political discourse: anything outside those limits—that is, opposed to the war— became by definition unpatriotic, seditious, even traitorous. Businessmen, therefore, could attack anything that affected war production as pro-German or worse and portray themselves as patriotic Americans. Because workers' demands and strikes did often affected production, unions found their legitimate grievances characterized as un-American or pro-German. In effect, the claims of lumber baron John Kirby—the BTW's hated nemesis—that unions were "treason, pure and simple" had become national policy.[2]

The AFL responded by acquiescing to and becoming part of the war effort. The AFL possessed stable and strong branches in many industries, so this may have been a pragmatic, if timid, decision. But the IWW lacked such strong unions and tended to organize in previously nonunionized industries, where employers generally harbored strong antilabor sentiments. Such industries could readily attack the IWW for both its antiwar stance and its advocacy of sabotage. This was particularly true of the oil industry, a critical part of the war effort. The appearance during the war of Wobbly organizers in Oklahoma—then the largest petroleum-producing state—gave industry officials the means to crush unionism and offered the federal government the opportunity to pursue the charges it hoped would destroy the IWW.[3]

State authorities, too, profited. Wartime patriotism finally allowed the business progressives in the Democratic Party to wrest control from the more populist, rural elements. Authorities also forged alliances with oil industry executives and with business-minded agricultural groups such as the Farm Bureau. To enforce its new power, the state government relied on local Councils of Defense, often dominated by oil company officers, and on the state press. This support, merged with vigilantism, allowed the opponents of socialism and radicalism to destroy an enemy they had not been able to eliminate in peacetime. But while the SP was the conservatives' long-term target, in the short term they focused their attacks on the IWW.[4]

Although the most vicious assaults followed the Green Corn Rebellion, the IWW knew what the April 1917 declaration of war on Germany implied: that anyone who opposed the war was unpatriotic at best and traitorous at worst. The passage of the Draft Act in May and the Espionage Act in June only confirmed the IWW's fears. The Espionage Act posed a particular threat as it called for $10,000 fines and imprisonment for up to twenty-years for obstructing the armed forces, fomenting insubordination or disloyalty, or disrupting recruitment or enlistment. The union's leadership realized that the definition of obstruction was broad enough to include virtually anything. The later Lever Act, which made it illegal to disrupt the flow of war-related material for any reason, including strikes, only complicated matters for labor as a whole and for the Wobblies in particular.[5]

Wobblies elected to play it safe and soft-pedal or abandon antiwar rhetoric. AWIU head Forrest Edwards ordered antiwar and antimilitary soapboxers off the streets. Editorials in *Solidarity* urged workers to concentrate only on organization. Over 95 percent of Wobblies registered for the draft, and most served when called up. Within the GEB, Haywood held off the demands of Frank Little and others who wanted the IWW to take a firm stand in support of the union's 1916 declaration against war.[6]

Oklahoma Wobblies mostly agreed with Haywood and Edwards. Soapboxers disappeared from the streets of Drumright and other oil towns. Writing from Oklahoma City, organizer M. A. Hathaway told Arthur Boose, then in Tulsa, that suspending street-speaking was a good idea because job delegates were more effective in organizing than in propagandizing. Boose himself urged members to take jobs with their "cards shelved and buttons hid." But the IWW's cautious position did not protect it from attack. In Oklahoma especially, the Wobblies would face the assault of federal spies, locally organized Home Guards and Councils of Defense, and the oil companies' private police.[7]

This opposition was not new. Many self-styled "Progressives" considered the Wobblies' economic program and philosophy treasonous. Viewing themselves as morally superior, these men sought to rationalize politics, protect the independent property owner, and restore "harmonious" relations between employer and employee. Many worried that labor militancy was percolating down to the unskilled and semiskilled workers and meant that they would join unions. The war offered the Progressives a chance to use the martial spirit to create a disciplined, rational, and moral America, a goal threatened by the IWW and the Left. They found in President Woodrow Wilson an ally in their efforts to eliminate that threat.[8]

As early as 1915, the Wilson administration planned to suppress the IWW based on the assumption that the union had engaged in an interstate conspiracy to create disorder. But investigations found no such criminal plots. Attempts to use alien deportation laws also failed. First, the Wobblies were not intimidated by the law; second, most of the members and leadership were native born; and third, many employers feared labor shortages and only wanted the IWW silenced, not removed from the labor market.[9]

The escalation of the war gave the federal government new tools to deal with radicalism. In August 1916, Congress authorized the creation of a national council as well as state Councils of Defense under the National Defense Act, although organization of the national council was not fully implemented until March 1917. That same month, U.S. Attorney General Thomas Gregory approved the creation of the American Protective League (APL), a volunteer citizens' organization directed by the Bureau of Investigation. Both organizations—as well as the military—were to engage in surveillance of radicals and whip up public support for the war. The passage of the Espionage Act added another dimension to the crazy-quilt strategy to suppress dissent. By July 1917 the government planned to use the act to indict the IWW leadership on conspiracy charges, even though acts by individual Wobblies were not illegal. Attorney General Gregory showed some reluctance at first to use this new tactic but reversed his position on July 31, only days before the Green Corn Rebellion.[10]

Oklahoma authorities eagerly jumped on the bandwagon, aware that support for the war was not strong among many state residents. Although the state legislature had adjourned before approving a state defense council, Governor Robert L. Williams ordered the creation of one, with associated county chapters. These local councils dominated the home front in Oklahoma and showed enthusiasm for counterinsurgency. The strongest local chapter was, not surprisingly, that of Tulsa County. It was controlled by oil company executives and included Tulsa City Attorney John B. Meserve (a particular foe of the IWW). It also had the support of local newspapers. APL chapters also appeared, with a Tulsa chapter—again, led by oilmen—founded on May 26, 1917.[11]

Perhaps worried that forming the defense councils and the APL would take too long, state oil companies had already organized their own security forces and hired private detectives to act as labor spies. At least three spies from the Burns, the Pinkerton, and the Kirk and Gustafson agencies had joined the Tulsa IWW local, and two became union officers. One Pinkerton agent assured federal authorities that Wobblies who committed sabotage were fired from jobs. Carter Oil, a Standard of Indiana subsidiary, went further than just using spies. It had a private police force led by H. H. Townsend, a former Tulsa assistant police chief. Most oil companies readily cooperated with the APL and the justice department. For example, Fred Cook, the Tulsa head of Prairie Oil and Pipeline Company, regularly provided his spies' information to John Whalen, the local Bureau of Investigation agent, and to the U.S. attorney's office.[12]

From the start, the IWW realized that the federal government was now cooperating with state authorities, local businessmen, and national corporations in a virtual conspiracy to crush the union. The leadership knew that undercover agents had infiltrated the locals. The IWW's dilemma involved balancing its need to identify those agent provocateurs against its organizing activities. Yet the Wobblies were surprisingly optimistic, almost naive, in light of the dangers awaiting them. Much of the IWW's confidence may have stemmed from two factors: first, the success of the 1916 campaign; and second, the knowledge that the draft would likely produce a labor shortage that would drive up wages, especially in the harvest. In addition, the Wobblies' plan of avoiding the war issue and sticking just to organizing may have convinced the leadership, however wrongly, that they would be left alone. Certainly the refusal of the AWIU to affiliate with the WCU suggests this, as does its oral agreement with the Non-Partisan League, a North Dakota–based farmers' organization, to try to raise wages to five dollars a day.[13]

Initially, the optimism may have seemed justified. The state's anticipated 32-million-bushel winter wheat crop was about 31 percent of the national total, and local officials feared a shortage of hands. The free employment bureau office at Enid estimated a need for five thousand harvesters but had received few applications. With reports suggesting a shortage of two million farm laborers (due to conscription), state authorities seriously considered importing a thousand German prisoners of war from France to work the harvest.[14]

Conditions favored the IWW. Before the harvest started, police at Cherokee released twenty-three Wobblies arrested for vagrancy after local merchants worried that the arrests might scare away help. Once the harvest started at Enid, *Solidarity* claimed that membership was up, as were wages. Expecting the Kansas harvest to be as large, the AWIU divided that state into five districts to aid organizing and hired an additional stenographer to keep up with the correspondence and record keeping. Job delegate J. R. Parker said southern Kansas farmers who refused to pay the union scale could not even obtain nonunion hands and had to travel forty to fifty miles to hire harvesters. "Even the Oklahoma, Louisiana, and Arkansas suitcase boys and scenery stiffs refuse to work anywhere around their locality," Parker reported.[15]

But other Oklahoma harvest events suggest that the IWW was overly optimistic. An atmosphere of repression permeated the state and the nation as a whole. The arrests of the WCU's leaders and of Wobblies in Drumright and Enid on charges of sedition and draft evasion should have warned the IWW. In addition, newspapers reported—without substantiation—that IWW members had planted railroad spikes in wheat fields near Alva to damage the reaping machines and had burned crops after police raided a Wobbly jungle.[16]

By late May, newspapers claimed an oil field general strike was set for June 5. The editor of the *Drumright Evening Derrick* said he believed the drillers and tool-dressers would not take part, but he acknowledged that the IWW had influence with the pipeline and ditching crews. Edward Sweeney, leader of the extralegal Patriotic Committee of Drumright Citizens, demanded the arrest and execution of "Benedict Arnolds" and draft resisters. Chief of Police Jack Ary of Drumright and Sheriff John Woofter of Creek County promised that agitators and pacifists would not be allowed to meet at city hall, and they deputized one hundred men as a special squad to quash any uprisings. A June 4 explosion at the Dewar water works blamed on the WCU only heightened tensions.[17]

Events elsewhere should also have warned the IWW. Strikes by lumberjacks in the Pacific Northwest had closed down 20 percent of all logging operations, while hard-rock miners struck for better wages and working conditions in Montana and Arizona. Although postwar investigator Alexander Bing found that the strikes were justified and the strikers had made reasonable requests, the presence of the IWW gave the lumber and mining corporations an excuse to break the strikes under the guise of patriotism. On June 9, Missouri National Guardsmen attacked the IWW headquarters in Kansas City and badly beat Wobbly Wencil Francik, who was later sent to Tulsa. City officials condoned the raid and added insult to injury by charging Francik with disturbing the peace. On July 12, vigilantes in Bisbee, Arizona, rounded up strikers at gunpoint, herded them onto boxcars, and dumped them in New Mexico's western desert. Finally, in late July, the main IWW headquarters in Chicago was burglarized, apparently by federal agents seeking evidence that the IWW had received gold from the German government.[18]

Within the union itself, Haywood and most of the GEB tried to stifle antidraft sentiment. Frank Little and Ralph Chaplin, editor of *Solidarity*, led a faction seeking open opposition to the war and the draft. But Haywood and the board majority argued direct opposition would play right into the capitalists' and authorities' hands. While the GEB agreed that the war was stupid and futile, the leadership contended only on-the-job organization could stop it. They ordered locals to stick to labor issues and to avoid alliances with any antiwar groups.[19]

Throughout the summer the OWIU did just that. To aid in recruitment, the GEB in June named Frank J. Gallagher traveling delegate for the Oklahoma and Kansas fields. The brawny, thirty-six-year-old Gallagher was a former AFL member and, unlike many other organizers, had several years experience in the Oklahoma oil fields. Believing he understood the labor situation better than did Phineas Eastman, an OWIU leader, he advocated a cautious approach. Gallagher, however, received little in the way of help from either the branch locals or headquarters.[20]

Still, the IWW thought it could benefit from the shortage of workers caused by the draft. In July, Tulsa branch secretary Frank Ryan wrote *Solidarity* and the *Industrial Worker* to say that a shortage of plant construction workers, tank builders, pipeliners, and ditchers offered an excellent recruitment opportunity. In August, an IWW-led strike at a Gypsy Oil pipeline camp near Drumright had raised pay to $3.50 a day and encouraged a non-IWW strike at a nearby Sinclair Oil camp. In contrast, five Wobblies who led a strike at Hominy in early August lost their jobs when the nonunion men returned to work.[21]

In fact, many oil workers increasingly encountered company hostility and a rising tide of jingoism. "Wobbly" and "IWW" had become convenient labels used to discredit legitimate labor demands, as AFL-affiliated Sinclair workers in Tulsa and Drumright learned when they struck for a 15-percent wage increase. They were accused of being Wobblies, and in response, the Tulsa County sheriff sent in special deputies to prevent any hostile demonstrations, while W. G. Ashton, Oklahoma labor commissioner, ordered Sheriff John Woofter to "crack down on the I.W.W. and prevent lawlessness no matter what measures you have to take." Despite the growing opposition of local authorities, the IWW was apparently also benefiting from the workers' unrest. By the end of August, military intelligence agent Charles Findley reported to his superiors that the OWIU now had four hundred men in Creek, Tulsa, Osage, and Okmulgee Counties.[22]

August 1917 proved a cruel month for the IWW. On August 1, vigilantes murdered Frank Little in Butte, Montana. Two days later the Green Corn Rebellion broke out. The hysteria the rebellion caused spread rapidly, fanned by the state press. The Tulsa plumbers' union local expelled two men suspected of being Wobblies. United States senators introduced three sedition bills aimed specifically at curbing the IWW, and President Wilson gave his personal approval to a scheme to raid several Wobbly halls simultaneously, including the one in Tulsa, in September. Even as the hostility grew, Phineas Eastman urged Frank Ryan to keep organizing in Tulsa, despite Little's murder as well as the shooting of a union member in Perry. But the fires of repression were growing hotter, stoked by the state's prowar press.[23]

Even before America entered the war, Oklahoma newspapers had grown jingoistic. As early as 1914 newspapers such as the *Muskogee Daily Phoenix* and Oklahoma City's *Daily Oklahoman* and *Oklahoma News* had attacked the IWW for opposing Wilson's intervention in Mexico and had supported the president's attacks on "hyphenated" Americans, especially those of German background. Once war was declared, the Oklahoma Press Association pledged its members' loyalty to the government's plans.[24]

While most of the state's press attacked the antiwar movement, the *Tulsa World* was the most pro-oil industry, prowar, racist, antiforeigner, and antilabor paper of them all. It focused much of its abuse on the IWW, which it repeatedly called "German bought and German controlled." Publisher Eugene Lorton, a staunchly conservative Republican, assigned staff members to keep up a stream of patriotic articles, while editorials by Glenn Congdon, the managing editor, openly called for violence against and even the murder of radicals. In an August 9, 1917, editorial, the *World* became the first paper to blame the Green Corn Rebellion on the IWW and charged the Wobblies with being German agents "seeking to cripple the energies of the nation by an attack from behind." Other editorials called the IWW anarchists who should be executed for treason and even accused Wobblies of stirring up "bloodthirsty Apaches" in Arizona.[25]

The Wobblies fought the editorials with gum-backed stickerettes known as "silent agitators." Soon a familiar sight in Tulsa, the eye-catching stickers urged workers to organize or "Slow Down. The hours are long, the pay is small, so take your time and buck 'em all." Others displayed the wooden shoe or the "sabo-tabby cat," both euphemisms for sabotage. Opponents said such stickers and flyers subverted the war effort. *Harlow's Weekly* printed an alleged IWW flyer that called for workers to join the IWW and refuse to be drafted. While intended to encourage workers, the stickers probably frightened people even more, further damaging the Wobblies' already precarious position.[26]

While the papers publicly vilified the IWW, the government moved in to stop the union. On September 5, Bureau of Investigation agents raided the main headquarters and forty-eight locals, including the Tulsa OWIU hall. The agents made no arrests at the Tulsa office but did seize ledgers, a song book, pictures of the local's members, IWW newspapers, pamphlets on sabotage, and four files of letters. The items constituted part of five tons of material given to a federal grand jury in Chicago, which on September 28 returned conspiracy indictments against 165 persons.[27]

All those indicted were charged with violating the Espionage Act by allegedly plotting to disrupt the draft and the military. Those IWW members with Oklahoma ties included Arthur Boose, Walter Reeder, Phineas Eastman, and Jack Law of the Oil Workers Industrial Union; Forrest Edwards of the AWIU; R. J. Bobba, an Arizona copper miner whose wife and family were in Oklahoma; and Harry Trotter, who had tried unsuccessfully to organize a Construction Workers Industrial Union local in Oklahoma City in the spring of 1917. Two non-Wobblies were also indicted: Henry Munson, WCU leader; and Stanley J. Clark, a lawyer and Socialist Party member.[28]

Either because he knew of the raids beforehand or because his wife was ill, Eastman had resigned his post on September 5 and remained at large until 1919, but Boose was arrested on September 28 when three federal agents attended one of his educational talks in the union hall. Boose later recalled he was charged with being a fugitive from the bogus Minnesota murder indictment and "with almost every crime I had ever heard of, including lack of what is commonly called patriotism." In late October, he and Walter Reeder were taken to Chicago in shackles.[29]

The raids and arrests completely altered the union's leadership. Many minimally qualified persons wound up in positions of authority. One such individual was Charles W. Anderson, a thirty-one-year-old Minneapolis shipping clerk who became AWIU and OWIU secretary-treasurer after Edwards's arrest. For reasons that remain unclear, he immediately decided to launch a full-scale organizing campaign in the Midcontinent. Anderson either chose the worst time to organize the oil fields or perhaps feared the union faced a life-or-death struggle.[30]

Favorable reports from Frank Gallagher and growing labor unrest in the Texas and Louisiana Gulf Coast oil fields may also have influenced Anderson. The Gulf Coast, AFL-affiliated Oil Field, Gas Well, and Refinery Workers had by fall already gained attention with the threat of a November 1 strike if its demands for union recognition, the eight-hour day, and a four-dollar daily wage were not met. In the October 6, 1917, issue of *Solidarity*, Gallagher said oil field jobs of alr kinds were plentiful in Oklahoma but added that Standard Oil was using a series of "independent" companies to break strikes and keep wages lagging behind production rates and profits. Conditions were good for organizing, Gallagher said, but a lack of job delegates handicapped the OWIU's efforts. He urged harvest workers to come to Oklahoma, take the jobs, and start organizing, but he also warned, "Get on the jobs and forget the brass band stuff and petty strikes and instead carry on a quiet system of education and agitation."[31]

Given those circumstances, Anderson may have simply believed the oil workers were ready to take action. To aid Gallagher, Anderson dispatched three men to Oklahoma: Wencil Francik, Michael Sapper, and Oscar E. Gordon. A Russian-born, thirty-five-year-old Iowan, Francik joined the IWW in 1909 and had recently been beaten during a raid on the Kansas City IWW local. Also Russian born, Sapper was thirty-two, had been a Wobbly since 1913, and was syphilitic. Thirty-one-year-old Gordon was a former army sergeant who joined the IWW in 1915. An excellent soapboxer, he had organized harvest hands in Kansas, Nebraska, and the Dakotas.[32]

Gallagher, suffering from some undisclosed illness, was displeased with his new help and believed their background in the strong-arm tactics of harvest organizing was ill-fitted for the oil fields. "The harvest worker was between fires, and was a stranger in a strange land, but the oil worker is on his own stamping ground and becomes antagonistic as soon as force is used," Gallagher warned Anderson. He reluctantly agreed to work with the men, but he wrote Anderson to say he was risking his own liberty and only took the job because he hoped it would prove worthwhile in the long run. Either to show his displeasure or because of his illness, Gallagher put off scheduled meetings with the men in Minneapolis and in Augusta,

Kansas. He instead went to Ardmore and asked to meet with them on November 8 in Arkansas City, Kansas, to discuss publicity and organizing.[33]

Gallagher's apprehension was understandable. In its issue of October 27, 1917, *Solidarity* had dismissed rumors of hostility in the oil fields and declared it safe to carry an IWW card in Oklahoma despite the arrest of a job delegate in Henryetta and police claims of an IWW dynamite plot in Oklahoma City. Two days later, on October 29, an explosion damaged the home of J. Edgar Pew, a manager for Carter Oil. The *Tulsa World*'s huge first-page headline cried, "I.W.W. Plot Breaks Prematurely in Blowing Up of Pew Residence." The paper asserted that the explosion was the result of a bomb that was part of a wide-ranging IWW conspiracy, and it called on 250 men to join the Tulsa Home Guard to counter the plot. Tulsa City Attorney John B. Meserve claimed that the Wobblies planned to destroy the city waterworks and place bombs in various neighborhoods.[34]

Four hours after the explosion, Tulsa police arrested oil worker W. J. Powers for questioning. Police chief E. L. Lucas claimed that the bombing was the first in a massive plot to destroy oil company property and the homes of oil company executives. The *World* claimed that Powers was caught trying to leave town and that his denials that he was an IWW member only proved he was one. The paper also claimed itself to be an IWW bomb target, citing four letters allegedly from the IWW that called for the "downfall of certain capitalist newspapers." On October 31, the paper began endorsing vigilante actions against political radicals. The *World*'s editorial writer suggested placing Wobblies either in concentration camps or lynching them. "[W]hat is hemp worth now, the long foot?" the editorial asked.[35]

Tensions remained high a week later when, on November 5, 1917, the Tulsa police, without a search warrant, raided the IWW offices on Brady Street. But while the two September raids had been carried out by federal agents, this November raid was a local effort intended to destroy the oil workers' union. In fact, agent John Whalen had urged Meserve and city officials not to raid the hall. When the police entered, local secretary E. M. Boyd was putting stamps in dues books. After he told police the union paid rent for the hall, Boyd asked to see a warrant. An officer replied that "he did not give a damn if we were paying rent for four places [as] they would search them whenever they felt like it."[36]

The police found no incriminating material but arrested the eleven men present on vagrancy complaints and jailed them. The vagrancy charge was patent nonsense, as Boyd was a salaried union officer and almost all the others held jobs. Significantly, one of the arrested men, John McCurry, was actually a Pinkerton detective working undercover while serving as the local's publicity agent. Using a pretense, officers later released McCurry from the cell.[37]

The *World* proclaimed war had been declared on the IWW, and a Tulsa police captain promised to arrest any men found loitering near the IWW offices. They backed up the claim with an additional arrest on November 6. Curiously, that man—identified as either E. G. Morris or E. G. Morrison—was not tried with the others, nor does it seem he was ever jailed. Based on evidence in federal surveillance records, it seems likely that Morris was a Kirk and Gustafson detective whom federal agents reported was elected to replace Boyd as OWIU local secretary.[38]

The eleven men arrested were brought to trial on November 8, 1917, before municipal judge T. D. Evans, later Tulsa's mayor during the city's infamous 1921 race riot. In a bizarre proceeding for a municipal vagrancy trial, the prosecution tried to paint the prisoners as "loafers" set on disrupting the war effort and attempted to tie them to the Pew bombing. Defense attorney Charles Richardson argued that city authorities failed to show any misconduct on the Wobblies' part and charged the city had harassed the men because the IWW was "the only fraternal society in the country which requires that every man, before being accepted, shall establish the fact that he is a bona fide worker and wage earner." After five hours, Judge Evans adjourned the trial until the next afternoon.[39]

Disappointed at the lack of sensational information from the trial, the *World* published a virulently anti-IWW editorial, probably written by Congdon. Entitled "Get Out the Hemp," the editorial openly called for lynching Wobblies and suggested that "a knowledge of how to tie a knot that will stick might come in handy in a few days." The editorial writer justified the remarks by saying the IWW was an enemy of the country and threatened the production of petroleum for the war. In order to defeat Germany, true patriotic Americans had "to strangle the I.W.W.'s. Kill 'em just as you would kill any other kind of snake. Don't scotch 'em; kill 'em. And kill 'em dead. It is no time to waste money on trials and continuances like that. All that is necessary is the evidence and a firing squad. Probably the carpenters' union will contribute the timber for the coffins."[40]

When the trial resumed on November 9, the prosecutors rested their case solely on the defendants' IWW membership. None of the men had criminal records, and police failed to tie them to the Pew bombing. In fact, one defendant—Swedish-born Gunnard Johnson—had been in Tulsa only three days, while another, Joe French, had just arrived on the day of the raid. The trial ended at 10:30 P.M., with Judge Evans finding Johnson guilty of the rather odd offense of not owning a Liberty Bond, a government security used to finance the war. The others had previously agreed that the decision in Johnson's case would apply to them all. Then, strangely, Evans ordered five defense witnesses arrested, tried them, and found them guilty of the same offense. Among the five were former OWIU local secretary Frank Ryan and W. H. Walton, a plumber expelled from the AFL local who was now an IWW job delegate. All the men were fined $100, although apparently none was asked to pay the fine. In justifying his actions, Evans said, "These are no ordinary times."[41]

Two hours later nine policemen took the sixteen prisoners from the city jail and placed them in cars, allegedly to transfer them to the county jail. Four blocks away, at a railway crossing, about fifty black-robed and hooded Knights of Liberty were waiting. "It was cold. You could see the frost on the railroad ties," Joe French later recalled. Mob members carrying rifles, shotguns, and revolvers commandeered the cars, clearly with police acquiescence. The prisoners were tied up, placed back in the cars, and, with guns at their heads, were driven three miles into the Osage Hills. There began a brutal charivari, as the mob stripped their victims to the waist, tied them to trees, and whipped them. The invocation "In the name of the outraged women and children of Belgium" accompanied each whip stroke. The attackers then applied hot tar and feathers from a slit pillow to the prisoners' backs. After being warned they would be killed if they returned to Tulsa, the prisoners were driven further into the hills, where the cars crashed into a barbed wire fence that lay hidden in their path. Abandoning their victims there, the vigilantes returned and afterward burned the men's shirts, jackets, and money.⁴²

That the mob was made up of prominent businessmen and had police cooperation seems indisputable. The police said they could not identify any mob member. But Boyd, in testimony to National Civil Liberties Bureau (NCLB) investigators, said the vigilantes provided police chief Lucas and a detective named Blaine with robes. An investigator hired by the IWW later reported the mob included J. Edgar Pew, former assistant police chief H. H. Townsend, and several police officers under indictment for killing a teenage boy. Also reported to be in the mob were members of the Tulsa Chamber of Commerce, a local minister, and F. W. LaFallette, business agent of the plumbers' union. The Tulsa press had also apparently received advanced notice of the assault. Glenn Congdon of the *Tulsa World* and his wife went along as spectators, although Congdon may have taken part as well. A *Tulsa Times* reporter said the Knights, resembling Ku Klux Klansmen, provided "a picturesque scene." A lieutenant in the Tulsa Home Guard later admitted that the Knights of Liberty obtained their weapons from the guard's arsenal, and Deputy U.S. Marshal John Moran told an NCLB investigator, "You would be surprised at the prominent men in this town who were in this mob."[43]

The day after the floggings, which the Wobblies soon dubbed the Tulsa Outrage, the police continued their crackdown on the IWW, promising to arrest members as soon as they were discovered. When one victim was arrested in nearby Sand Springs, Tulsa police told the local police chief to "turn him loose and tell him to keep going—away from Tulsa." The police also shut down the IWW hall, a fairly moot act. Most of the victims, fearing for their safety, had fled to the IWW local in Augusta to regroup, but several weeks after the Outrage Tulsa police rearrested one man after eight officers surrounded his Tulsa home and claimed he was carrying a pistol. Judge Evans admitted that the man, a longtime Tulsa resident and a member of the AFL carpenters' union, was not guilty, but still fined him $100. Police threatened to keep arresting him until he left Tulsa.[44]

The arrests and whippings inspired Creek County authorities, who had earlier arrested two Wobblies for soapboxing near the Drumright post office. On November 6 Drumright police and Creek County deputies invaded the local IWW hall, while federal agents on November 9 raided the offices of attorney Grace Arnold, who was sympathetic to the IWW and a friend of E. M. Boyd. In a second raid on November 10, police and deputies smashed windows and furniture, seized IWW literature, and arrested members for vagrancy. Officers forced local secretary Chester Macklin into a car and drove him out of town. Federal agents then arrested Macklin November 12 at Skedee, near Pawnee. In a letter to Charles Anderson, Michael Sapper said he had fled to Cushing after a death threat, but police there were stopping migratories and searching boarding houses for Wobblies. Armed gunmen also patrolled the oil fields, he said. Wencil Francik warned Anderson that it was unsafe to reopen the Tulsa hall.[45]

Attacks on the IWW continued in the press, especially after news of the Bolshevik Revolution in Russia reached the United States. Two days after the Outrage, the *World* praised the Knights of Liberty and called on every citizen to keep an eye on his neighbor, lest he show unpatriotic tendencies. The *Daily Oklahoman* called the Tulsa action high-handed, but declared that the IWW's treason was common knowledge; while the *Drumright Evening Derrick* declared agitators should be used "as a decoration on a telephone pole." The industry's *Oil and Gas Journal* praised the Knights of Liberty as a new Ku Klux Klan and, conveniently forgetting the strikes of the summer, claimed Oklahoma and Kansas oil workers were satisfied with conditions.[46]

In response, the IWW charged that the Knights of Liberty were part of an oil industry attack on all workers. The union found support from the Socialist Party and the newly organized National Civil Liberties Bureau, which investigated the incident. The *Oklahoma Federationist*, the official newspaper of the Oklahoma State Federation of Labor, gave the Wobblies some backhanded praise. Before the Outrage, the *Federationist* had supported the IWW's claims, although without admitting it, when it challenged oil company and newspaper assertions that Oklahoma oil workers were well-paid and satisfied. After the arrests of an organizer for the Amalgamated Meat Cutters and Butcher Workmen in Oklahoma City and an AFL oil worker in Drumright—both of whom were accused of being Wobblies—the newspaper declared that businessmen, whose opposition to unions had helped create the IWW in the first place, were now using the IWW to attack the labor movement as a whole. The *Federationist* reiterated the loyalty of the AFL to the war effort and reminded its members to avoid contact with the Wobblies.[47]

Fueled by the press, anti-IWW hysteria spread to other communities. On November 15, 1917, the Sapulpa police chief shot an alleged Wobbly who supposedly attacked him with a knife, while Okmulgee authorities arrested two men accused of being Wobblies and of possessing a "powerful alkaloid" poison for some unexplained antigovernment scheme. The IWW was accused of setting fire to a railroad station in Henryetta and to Tulsa's Mayo Hotel and were blamed for a storage tank explosion that killed two men near Cushing. In Cleveland County, a posse sought Wobblies accused of trying to torch a bridge west of Norman, and the Santa Fe depot at Moore, while Pontotoc County officials alleged the IWW planned to attack troop trains. When a train of the Saint Louis and San Francisco Railroad (commonly called the "Frisco") crashed on November 25, the November 26 *Drumright Daily News* declared "Frisco Flyer Speeds into I.W.W. Trap of Death: Three Killed." An investigation revealed the wreck was actually caused by a metal bar placed on the tracks by three young boys, but the *Daily News* declined to apologize, saying train wrecking was in harmony with IWW practices.[48]

Rumors and innuendo proved only slightly less important than federal indictments. On November 20 and 21, federal agents, assisted by local police, raided the OWIU locals in Augusta and El Dorado, Kansas, ostensibly to collect evidence against Phineas Eastman for the Chicago conspiracy trial. Federal officers seized newspapers, literature, and correspondence, while local authorities held forty-two men on "John Doe" warrants for vagrancy. Among those arrested were two members of a failed August pipeliners' strike in Skiatook: A. M. Blumberg, a rank-and-file member; and Albert Barr, a former secretary of the Tulsa local, who was nicknamed "Dr. Barr" because of the syringes he kept for his narcotics addiction. They were joined by other Wobblies with Oklahoma ties, including Carl Schnell, Joseph Gresbach, E. M. Boyd, Frank Gallagher, Wencil Francik, and Michael Sapper. Tulsa police arrested Francik on November 17, 1917, and reportedly had to protect him from an angry mob. Pawnee police arrested Sapper in late November, while Gallagher was picked up in Kansas City on December 1. E. M. Boyd almost avoided arrest until police raided the IWW hall in Saint Louis, Missouri, in March 1918. Phineas Eastman was arrested late in 1919. The men would all be among twenty-eight Wobblies selected for trial by federal authorities in Kansas.[49]

Arrests continued in Oklahoma as well, with Chickasha authorities arresting two alleged Wobblies—Harry "Nuf Said" Casey, an itinerant watchmaker, and L. W. Oliphant—in connection with the Pew bombing. Both men were later released, but the major break in that case came with the December 28, 1917, arrest of Charles Krieger. For Krieger, a former Philadelphia machinist who joined the IWW in the oil fields, it marked the beginning of more than two years of captivity without a trial. Federal authorities also levied charges of espionage and antidraft sedition against Drumright OWIU branch secretaries Chester Macklin and Fred Edgecombe, although both were later acquitted. With these arrests, the IWW's opponents shifted their strategy from the blunt force of vigilantism to the subtler tactics of the judiciary.[50]

The government's use of the federal courts in its assault on the IWW is not surprising. Federal judges had issued injunctions and presided over conspiracy trials intended to cripple the labor movement since the 1890s. Such judicial onslaughts helped turn the AFL from militancy and toward "business" unionism and shop floor economic issues, while the courts' treatment of the Western Federation of Miners at the turn of the century was a factor in the IWW's rejection of political action. But in the Kansas and Chicago trials the courts would do more than merely restrict labor activities. They would effectively outlaw the IWW.[51]

The first of these legal assaults was the well-known Chicago conspiracy trial, which began in May 1918. Although 165 persons were originally named, several had their indictments quashed or dropped. Included in that group were Walter Reeder, later rearrested and convicted in the Universal Union case; Henry Munson, already convicted in the WCU cases; and Harry Trotter, released during the trial without explanation by Judge Kennesaw Mountain Landis. The others—with the exception of Phineas Eastman, then a fugitive—endured a five-month long trial that ended in August with their convictions. The testimony of Frank Wermke—a former Wobbly and alleged "gun-thug," or armed robber, who had joined the army to avoid prosecution—that he had placed railroad spikes in Oklahoma wheat fields to damage farm machinery proved important to the prosecution. While the defense refuted many of Wermke's allegations, the jury believed him and convicted the defendants. Of the Oklahomans, Arthur Boose received a five-year sentence, Stanley J. Clark was sentenced to ten years, and R. J. Bobba received a one-year sentence. All three were fined $30,000 each.[52]

Clark faced the most stringent prosecution attacks, perhaps because the case against him—for making antidraft speeches in Arizona during the Bisbee and Jerome strikes—was weak. A radical Oklahoma SP member who had supported Wilson in 1916, he had been an IWW member for two months in 1905, and then simply as a formality to aid fund-raising. From 1912 until his arrest, Clark worked as a lawyer and was therefore ineligible for IWW membership. Although the prosecution hoped to portray Clark as seditious, defense witnesses testified that Clark had asserted the draft's constitutionality and told audiences only the soldiers and workers themselves could stop the war.[53]

Despite the overall importance of the Chicago trial, the Kansas indictments carried greater significance for the IWW, as they affected the industries that were the financial and membership lifeblood of the union. At first the Department of Justice hoped that W. P. McGinnis, the U.S. attorney for Oklahoma's eastern district, would pursue indictments of the IWW, an expectation based on McGinnis's zealous prosecution of the WCU. But McGinnis, unlike his superiors, apparently believed the IWW was a legitimate labor organization working for better wages. When oil company executives demanded that federal agents arrest and detain all the Wobblies in the Midcontinent fields, McGinnis refused, saying he could not have IWW members arrested merely on suspicion of what they might do. McGinnis could resist such demands because extralegal attacks and local raids had driven most Wobblies into Kansas. As a result, federal authorities turned to McGinnis's counterpart in Kansas, Fred Robertson.[54]

Less zealous than McGinnis, Robertson still played his role with diligence, even while conceding that the federal case was weak. Nevertheless, in March 1918 a Wichita grand jury indicted twenty-eight Wobblies on charges of conspiring to violate the Espionage Act and Lever Act by interrupting the flow of petroleum. The government delayed their trial until after the Chicago convictions; then Assistant U.S. Attorney General Claude Porter prepared a new indictment based on those same charges, but legal flaws caused it to be quashed. In June 1919 a special grand jury returned a third and final four-count indictment that treated the IWW as an outlaw organization that had conspired to avoid the draft, encouraged disloyalty in the military, and violated the Lever Act by restricting food and petroleum production.[55]

But what was really criminal were the appalling Kansas jail conditions the men endured for almost eighteen months. The overcrowded cells were unventilated and dank with moisture and sewage. Many prisoners—including Francik, Gresbach, and Blumberg—developed tuberculosis, pneumonia, scarlet fever, typhoid, and influenza. One prisoner died, while two others went insane. While their jail treatment was terrible, the prisoners' luck worsened when their December 1919 trial in Kansas City, Kansas, coincided with both a nationwide coal strike and the height of the Red Scare.[56]

The evidence against the IWW consisted mostly of the organization's newspapers and pamphlets and statements from assorted witnesses, including Frank Wermke, who repeated his Chicago testimony. Little evidence directly involved the defendants, although in one bizarre twist Oklahoma governor James A. B. Robertson—who had served on Oklahoma County's draft board—testified that Michael Sapper had registered for the draft in Oklahoma City, which clearly repudiated the charge of draft obstruction. But the union, which was near bankruptcy, lacked the funds needed to bring in its own defense witnesses. Instead, defense attorneys Caroline Lowe, George Vanderveer, Fred Moore, and Otto Christensen had to rely on their summation to show the inadequacies of the government's case.[57]

Despite its best efforts, the defense was unable to persuade the jury, who seemed especially swayed by exaggerated tales of Wobbly violence from federal agent Thomas J. Howe, an agent provocateur in the IWW since 1916. On December 18, 1918, the jury returned guilty verdicts against twenty-six defendants, a twenty-seventh having changed his plea to guilty and another remaining at large. As head of the AWIU-OWIU, Charles W. Anderson received the longest sentence, nine years. The Oklahoma Wobblies were sentenced as follows: Frank Gallagher, eight years; Phineas Eastman (who was arrested in time for the trial), Oscar Gordon, Michael Sapper, and Wencil Francik, seven and one-half years; Albert Barr, A. M. Blumberg, and E. M. Boyd, five years; Joseph Gresbach, three and one-half years; Carl Schnell, three years. On December 19, the men were taken to Leavenworth federal penitentiary, where they joined the Chicago IWW defendants and members of the Working Class Union and Universal Union.[58]

Legal appeals to free the men began almost immediately, despite the financial strain more legal costs placed on the IWW. On June 12, 1921, the U.S. Eighth Circuit Court of Appeals threw out the first count—that of a conspiracy to use force against the government—in the Wichita indictments, which freed twenty-one of the men, including Schnell, Gresbach, Boyd, Barr, and Blumberg. The other six—Anderson, Eastman, Francik, Gallagher, Gordon, and Sapper—remained in jail as the longest portion of their sentences was under the third count of obstructing the draft and the military. Support also came from other quarters. A Socialist-sponsored "Children's Crusade" aided the release of several prisoners, and by 1923 Oklahoma's newly elected U.S. Senator John Harreld, a

Republican, and Governor John C. Walton, a Democrat, called for freedom for all IWW and WCU prisoners.[59]

Ironically, gaining freedom eventually created an internal crisis for the IWW. While Oscar Gordon gained his liberty on a legal technicality, other Wobbly prisoners accepted conditional amnesties. But Wencil Francik joined with twenty-seven other Wobblies convicted in other trials in refusing the amnesties, which required the men to agree to be law-abiding; and federal authorities refused to offer Frank Gallagher an amnesty, deeming him too dangerous. The thirty remained jailed until President Calvin Coolidge pardoned them on December 15, 1923. The issue of accepting the amnesties remained, however, and soon fueled a devastating schism in the IWW.[60]

In the long run, the repression of the IWW was more than simple wartime hysteria. National mobilization fit into the needs of the incorporating forces, such as Oklahoma's business progressives; and government and business viewed labor's demands, even reasonable ones, as detrimental to the war effort. To achieve their goals the incorporationists used legal means such as the courts, quasi-legal entities such as the American Protective League and the Oklahoma Councils of Defense, and extralegal devices such as vigilantism and covert domestic espionage.

In Oklahoma, the war gave the business-oriented forces the opportunity to defeat all their opponents, including conservative agrarians, radical Socialists, and even moderate labor organizations. The state's incorporationists believed their economic position precarious, despite their control of land and crop prices, the banks, and local politics. They had somewhat less power at the state government level, but all Oklahoma governors did have ties to powerful corporations. Still, the business progressives profited from growing rifts among labor, radicals, and the conservative populists. Workers and tenants—products of incorporation—were becoming the majority, while farmers were as quickly becoming a minority.

Initially united in common cause against incorporation, these groups rapidly moved in different directions.[61]

In Oklahoma, the business progressives exploited the split by emphasizing economic development and growth as the solutions to state problems. In this way they were able to unite the state government and "enlightened" capital. In addition, this active, coordinated capitalist force saw how to draw farmers closer to its interests and away from labor's, especially as wartime demands for cotton, wheat, and oil brought new wealth to the state. More successful farmers came to share the businessman's view of organized labor as a threat. Oil money further increased corporate control because it undermined local merchants and industry and destroyed the relationship between producers, small business, and labor.[62]

But, most important, the war allowed the business community, supported by some agricultural elements, to use the issue of patriotism against labor and radicalism, and especially against the IWW. With a violent image created by the probusiness press and furthered by intemperate statements from individual union members, the Wobblies made an easy target; and hostility toward the IWW could then be used against other radical groups such as the Socialists. The Green Corn Rebellion proved a windfall for the incorporators, allowing them to demonize the Wobblies as the source of all wartime social problems. Further, in trying to unionize the petroleum industry, the IWW in Oklahoma was up against huge national corporations and their local counterparts, which controlled virtually every aspect of life.

The alliance between business and government during the war only worsened things for the IWW. The Oklahoma State Council of Defense was allowed to define the limits of dissent, and its definitions placed all radicals beyond the pale. The Wobblies' earlier antiwar position, structural and financial problems, and contempt for institutions such as the courts and police kept the union on the defensive, while the federal government's successful assault on the IWW leadership left the

Wobblies essentially impotent and unable to respond effectively. The severe sentences given the Wichita Wobblies also served to warn others of the fate awaiting them if they criticized the new order. Yet, amazingly, the IWW did survive the onslaughts, the trials, and the lynch mobs to return to Oklahoma. But the authorities were waiting, armed with a criminal syndicalism law and backed by judges who were, with one exception, eager to jail radicals.

6

Trials and Tribulations
The IWW and the Red Scare in Oklahoma

ATTACKS on the IWW and other radicals in Oklahoma continued through 1918 and 1919. Mobs assaulted Wobblies at Tahlequah, Tulsa, and Durant in April 1918, and newly formed Ku Klux Klan units began patrolling military bases to drive out bootleggers, prostitutes, and laborers "infected with the I.W.W. spirit." The state press continued to claim that German gold funded the IWW and printed wild rumors about IWW plots to seize massive amounts of food and to dynamite oil refineries or whole towns in revenge for the Tulsa Outrage. The *Daily Oklahoman* called radicals and antiwar agitators "defectives" and repeated unsubstantiated stories of Wobbly crop burning and machine wrecking. The *Tulsa World* urged the deportation of foreigners who criticized the government and called for stern action against "those ungrateful sons of native sires who have absorbed the political and economic fallacies of the foreign disturbers."[1]

School officials, too, feared Wobbly infiltration. Having already outlawed teaching the German language in elementary schools, school districts began ferreting out IWW influences. The state school superintendent revoked the teaching certificate of Katherine Bondhauer because she had used an IWW songbook in her Harper County classroom, while in Healdton students burned a similar songbook as they sang "America" and "The Star-Spangled Banner." Fears remained so rampant that in late 1919, the president of the state Chamber of Commerce urged raising teachers' salaries because some educators had joined labor unions.[2]

Both the government and the oil companies also maintained covert surveillance. The Mid-Continent Oil and Gas Association hired operatives from the Thiel Detective Agency, one of whom served as a job delegate and collected funds for the IWW defense fund. Federal authorities also used spies, though occasionally the agents befriended the persons on whom they spying. At a Sand Springs glass plant, an operative told Wobbly Martin Englehardt he would be safe if he refrained from job agitation and handing out IWW literature. But the end of the war meant the military cut back on its own intelligence activities and relied on local police, the newly formed FBI, and reserve and recruiting officers.[3]

While federal action decreased, state authorities turned to the use of trials and restrictive legislation. The Oklahoma legislature approved several antisedition ordinances, banned the flying of any red flag (which ironically made it illegal to fly the state flag of that time), and considered a law to punish trespassing on or causing injury to oil company property, the latter aimed particularly at the IWW. But the state legislature's most important action was the passage of a criminal syndicalism bill in early 1919 that made it a crime to belong to the IWW or to possess IWW literature. This bill and the others marked an important change in dealing with radicalism. Strengthened by the war, the forces of incorporation gained complete control of virtually all political and judicial institutions. Now the business progressives and their allies could move away from the questionable legitimacy of vigilantism and toward the courts with all their legal niceties.[4]

The importance of this shift cannot be overestimated. Labor had long faced a hostile, probusiness court system in the east, but western courts often proved less favorable for business. With the appearance of criminal syndicalism laws in western states such as Oklahoma, incorporating forces gained the upper hand and, with it, the ability to define the acceptable boundaries of dissent. The trials that soon followed served to consolidate this newly achieved power.

English historian E. P. Thompson has observed that the law itself exists as a superstructure that adapts itself to the infrastructure of productive forces and social relations. As the de facto instrument of the ruling class, the law creates a set of rules that conform to the rulers' own rhetoric and ideology. Most important, the law defines what is legal and illegal, what is criminal and what was not. As a result, court cases emerge not just from disputes of the propertied versus the propertyless, but from alternative and conflicting definitions of what constitutes rights and property. The trials produced by such disputes become rituals to legitimize the power of the rulers—in this case, the Oklahoma business progressives.[5]

The IWW understood this. Some Wobblies saw the law as a sham to be exposed, and they used trials as showcases for publicity—such as following the Lawrence strike in 1912 and during the massive Chicago federal conspiracy trial in 1918. Other IWW members refused to recognize the authority of the courts, like those Wobblies who denied the authority of the court to try them and silently refused to participate in their own defense in the 1918 federal conspiracy trial in Sacramento. But in challenging the law, Oklahoma IWW members willingly participated in their trials and accepted the mechanisms of the court system. This choice may have been naive, but union members may have recognized that the jurors' natural sense of justice often overrode the letter of the law. Despite some convictions, this approach proved more successful in revealing the class-based nature of the prosecutions and in forcing the authorities to abide by their own rules and confront their own partisan and unjust actions.[6]

But in 1918 and 1919 the Wobblies could barely oppose either the courts or the vigilantes. The mass federal arrests and trials forced the IWW in early 1918 set up a General Defense Committee, which included Jack Law, former secretary of the Tulsa oil workers. The committee raised funds for the defense of IWW members, especially at the massive federal trials. But in 1919 the committee collected contributions for the defense of two

Wobblies arrested in Oklahoma: Grover Jackson "Jack" Terrell, the first person charged under the new criminal syndicalism law, and Charles Krieger, still awaiting trial in Tulsa for the Pew bombing. Krieger's case in particular would demonstrate how far the courts could be used to legitimize the image of Wobblies as violence-prone outcasts from decent society and to what extent prosecutors would violate their own laws to do so.[7]

Despite the Tulsa Outrage, federal agents and city police apparently believed they had yet to arrest the persons who had blown up Edgar Pew's home. The arrests of metal miner Harry Lyons in Bartlesville in November and of Chickasha watchmaker Harry Casey in December both came to naught. In Casey's instance, a convicted Texas murderer initially alleged that Casey had built the bomb in Lawton, but federal agents found the story far-fetched. Casey was later released.[8]

By that time federal marshals had arrested Charles Krieger. A journeyman machinist from Philadelphia, Krieger had worked at the Baldwin Locomotive Works there and was a member of the International Association of Machinists. In early 1917, he took a job in the copper mines in Prescott, Arizona, where he joined both the IWW and the Western Federation of Miners. He served as an IWW job delegate until the mining strikes of that summer. In August 1917, he came to Oklahoma and found work with Mount Copper Boiler and Iron Works in Bristow, where he remained until December 24, 1917. As testimony revealed, Krieger had been on a job in Perry, ninety-four miles from Tulsa, at the time of the Pew bombing. When arrested on December 28, he was waiting for a train to Saint Louis, on his way to visit his parents. Before he had left his job, Krieger asked another worker to join the IWW, and that man had turned him in to the police.[9]

The Krieger trial is especially important in IWW history as it is the only case in which a member of the union faced direct charges involving a specific incident of violence. By tying the IWW to the Pew bombing, Oklahoma authorities clearly attempted to prove what had long been claimed: that the Wobblies were dangerous anarchists bent solely on the destruction of the property and lives of law-abiding citizens. Lacking tangible evidence, the prosecutors constructed a scenario based exclusively on perjured testimony. But the prosecution ran up against a jury not entirely friendly to large oil companies and a judge who demanded that the trial conform to proper legal standards. The result was a rare legal victory for the IWW, especially notable in the hostile postwar period.

Krieger, who never denied his IWW membership, faced a preliminary hearing on January 5, 1918. He was bound over for trial for possessing allegedly seditious literature, hindering the prosecution of the war, and obstructing the draft, although it is not clear if he was charged with the Pew bombing at this time. Taken to federal jail in Muskogee, Krieger joined fellow Wobblies J. E. Wiggins, W. B. Montgomery, and Wencil Francik. While there he took part in a riot over prison conditions that ended only when guards used fire hoses on the prisoners. As punishment, Krieger and Francik found themselves placed in solitary confinement.[10]

How Krieger became a suspect in the Pew case is something of a mystery. Eugene Lyons, who covered the case for the Workers Defense Union, suggested that Carter Oil officials believed a Wobbly had blown up the Pew home. While four Wobblies were in jail at Muskogee, just Krieger and Wencil Francik had been near Tulsa, and only Krieger had passed through the city on October 25, 1917, on the way to his job in Perry. "Besides, he had a fine personality, was talkative, and could be paraded as the brain of the plot," Lyons wrote.[11]

The key figure linking Krieger to the bombings was Hubert Vowells, a convicted felon whom Krieger had met while both were jailed at Muskogee. Vowells, with two other men, was originally arrested on November 4, 1917, near Tulsa's red-light district for shooting street lamps while drunk. When confronted, he threw a bottle of nitroglycerine at a detective, who shot Vowells in the left leg. Vowells and the other men, John Hall and J. T. Foster, were first suspected in a train station robbery, but in late December federal agent John Whalen and Carter Oil detective J. W. Robinson interviewed Vowells regarding Harry Casey. Vowells offered little information other than that he was born in Owensboro, Kentucky, in 1888, was married, and had worked in Minnesota, Montana, Texas, and Mexico. He never mentioned Krieger in this interview, but evidence linking Vowells to the train station robbery may have convinced him that accusing Krieger might get him a shorter sentence.[12]

To corroborate Vowells's story, Carter Oil in March 1917 arranged to have an operative named George Harper housed with Krieger. A strikebreaker and a Pinkerton detective, Harper was also a convicted bank robber who turned to detective work only after losing a leg in a shootout with police. Placed in Krieger's cell, Harper later returned with reports that Krieger had confessed to hiring Vowells. Based on this information a federal grand jury indicted Krieger, Vowells, and Foster in July 1918.[13]

Vowells pleaded guilty and received a five-year sentence, but federal authorities dropped charges against Krieger and Foster, apparently convinced of both men's innocence. Krieger's release clearly upset Carter Oil officials, and they arranged for Robinson to rearrest Krieger as he left the Muskogee federal jail on October 16, 1918. Robinson took Krieger to Tulsa, where a complaint against Krieger was filed in district court on October 20. Krieger was denied counsel during his October 27 preliminary hearing, at which Edgar Pew testified he could smell nitroglycerine after the explosion; H. H. Townsend—head of security for Carter Oil—outlined the bombing theory; and George Harper reiterated his story about Krieger's confession, claiming Krieger had committed bank robberies in Arizona and was sent by a "sect of rebels" from Bisbee to commit violence.[14]

Conducting his own cross-examination, Krieger managed to demonstrate that two witnesses, David and Sadie Griffin, had never seen him with Vowells, Foster, or Hall, and that he knew none of the other men. Despite the hearsay nature of the testimony, the judge bound Krieger over for trial. In an unusual act of kindness, the court reporter provided a transcript, without payment, for Krieger's use once he found an attorney. A conservative Snyder attorney named R. H. Towne first represented Krieger, but a new defense team replaced Towne in January 1919. Leading the defense was the colorful Fred Moore. He was joined by Caroline Lowe, a well-known SP member, and George Bonstein, a poverty-ridden Tulsa lawyer.[15]

A Socialist Party lecturer and former school teacher, Canadian-born Lowe served as general correspondent for the SP's Women's National Committee. She had first supported the SP's 1912 anti-sabotage amendment, but her arrest during the 1913 Kansas City free-speech fight seems to have changed her opinion. She studied law at the Socialist Party's Fort Scott, Kansas, People's College, passed the Kansas bar in 1916, and by the same year was part of the IWW's legal defense team, serving with Moore in Everett, Washington, in 1916 and at the Chicago trial of 1918 with George Vanderveer. She and Moore were also serving as attorneys for the Wichita Wobblies.[16]

The two were a study in contrasts. Lowe was seen as sweet and spinsterly, while Moore had a reputation as a nonconformist who wore a broad-brimmed hat and cowboy boots and carried a handgun. Born in Detroit in 1882, he had grown up in Washington and had studied law at the University of Michigan before apprenticing himself to a Seattle attorney and taking the bar exam in 1906. Moore started out as a railroad attorney but suddenly shifted to defending labor agitators. Besides the Everett and Wichita trials, he had served as attorney for Wobblies involved in free-speech fights in Spokane (from 1909 to 1910) and San Diego (in 1912) and had defended Joseph Ettor and Arturo Giovannitti when they were accused of murder in the wake of the 1912 Lawrence strike. His greatest fame, however, would come when he defended Nicola Sacco and Bartolomeo Vanzetti and turned their case into a cause célèbre.[17]

But Moore was also an emotionally unpredictable man who too often argued with defense committees or his clients, suffered a series of psychological problems, and even simply vanished for weeks on end. His December 1917 disappearance led the IWW leadership to fear Moore had been kidnapped by federal agents, but he had, in fact, suffered a nervous breakdown. This forced the union to hire new attorneys to handle the Chicago and other trials. While many people thought Moore was simply unstable, his colleague Caroline Lowe explained that Moore suffered from a brain disorder that could produce periods of disorientation, although his reported addiction to cocaine probably added to his problems.[18]

In any case, Krieger had found the best lawyers for his defense. He was also fortunate that his case was assigned to the conscientious and fair-minded Judge Redmond S. Cole. The Missouri-born Cole, then twenty-eight, came to Oklahoma in 1909 after attending the University of Missouri law school. Settling in Pawnee, he soon served as county attorney and as mayor. From 1917 to 1919 Cole, a Wilson Democrat, was Assistant U.S. Attorney for Oklahoma's western district and took part in the prosecution of the "Jones Family," the Universal Union, and other radicals. He became a justice in Oklahoma's twenty-first judicial district, which included Tulsa, in 1919. Though mostly unsympathetic to political radicals, Cole possessed a genuine commitment to the principles of equal justice and a fair trial.[19]

On February 25, 1919, Cole quashed the first indictment against Krieger, saying that the state's case was based on utter hearsay and that none of the testimony from the preliminary hearing implicated Krieger. But the ink was hardly dry on Cole's order when Carter Oil's attorneys, led by Austin Flint Moss, swore out a new warrant before another judge the next day. By then it became clear the oil company was pursuing a vendetta against Krieger, perhaps as part of a plan to discourage further unionization in the oil fields. The *Oklahoma Leader* even noted rumors that Carter Oil itself had destroyed the Pew home as an excuse to crush the IWW, while a Tulsa attorney told Krieger the entire case was built on perjury: "They have gone to jails and offered immunity from every crime to every S. of a B. there, if he would testify just to a word against Krieger."[20]

On February 28, 1919, again based on testimony from Harper and Vowells, Krieger was bound over for trial on the felony charge of destroying a building with explosives, making him the first Wobbly actually charged with a violent offense related to the popular conception of IWW sabotage. Caroline Lowe told the press the accusation was the most serious charge ever brought against any IWW member: "You can see that this conviction would be a very bad black eye for the Industrial Workers of the World."[21]

Trial was initially set for September 22 but was delayed until October 6 while Moore dealt with the Wichita proceedings. Meanwhile, events outside the courtroom were hardly beneficial to Krieger's case. That fall steelworkers nationwide began a massive strike, as did Boston policemen. The summer had also witnessed the Seattle and Winnipeg general strikes and the start of antiradical raids ordered by Attorney General A. Mitchell Palmer. In Oklahoma, the press turned once again to fears of radicalism, the IWW, and the new bogeyman—Bolshevism. State authorities saw the hand of the IWW when a Drumright telephone operators' strike against Southwestern Bell erupted into a riot on September 22 after strike supporters marched on city hall. Governor Robertson sent in six companies of the National Guard, who arrived to find a relatively quiet situation. Police arrested eighteen persons, none of them Wobblies, on charges of inciting to riot, but the cases were later dropped. Robertson also called out the National Guard during the UMW national coal strike of 1919, which occurred while Krieger's trial was underway. The governor reacted in part to reports from operatives and private investigators claiming—erroneously—that the IWW was behind the strike in Oklahoma. Calling the strike a lawless conspiracy, Robertson on December 4 declared martial law in the coal fields.[22]

Faced with defending himself against perjured testimony, Krieger would have had difficulty enough finding a Tulsa

County jury willing to acquit him. With the unrest feeding public hysteria and tainting opinion against him, he might have thought the task almost impossible. Fred Moore believed that it was indeed impossible and requested a change of venue to Pawnee, which Judge Cole refused. But problems appeared almost immediately. First, Cole discovered the jury commission had failed to include anyone from Sand Springs, an area with a large working-class population and sizable Socialist electoral support. A lack of veniremen delayed jury selection, and when it did begin, twelve potential jurors were dismissed after they admitted they were prejudiced against the IWW. Selection was almost completed when on October 17 two jurors admitted having discussed the case two weeks earlier with a man who turned out to be Wobbly E. M. Boyd. A jury plus one alternate was finally seated on October 22, composed of a banker, a bank cashier, an oil producer, a building contractor, and nine farmers, one of whom was also an oil worker.[23]

When the trial began on October 23 it was an odd spectacle. Reporters from as far away as Seattle covered the case, while men carrying holstered revolvers came and left unrestrained. Eugene Lyons noted Cole smoked a cigar during part of the proceedings despite a No Smoking sign in the courtroom, and Cole himself wrote his wife that Moss seemed somewhat inebriated at one point, which had forced a brief adjournment. In the same letter, Cole noted that most of the prosecution's witnesses had served prison time.[24]

Although County Attorney Thomas L. Munroe was the nominal prosecutor, the case was actually being run by Moss and another Carter Oil attorney named L. G. Owens. According to Lyons, this so angered Munroe that he occasionally gave Moore tips regarding Moss's strategy. Cole, too, dropped hints to the defense counsels, while the court stenographer warned them that vigilantes had planned to run Moore and the others out of town. Later, Lyons learned that a local independent oil producer with a grudge against Standard Oil had stationed his own gunmen in the courtroom to protect Krieger and his attorneys.[25]

The prosecution, lacking any physical evidence to prove a bombing, instead relied mostly on testimony that was suspect at best. A crowded courtroom first heard George Harper testify that Krieger had said he had paid John Hall $400 to hire men to destroy the Pew home and also to kill John D. Rockefeller. Hall, in turn, had hired Vowells and a man named George Benson, who were each to receive $125. Under questioning, Harper admitted having served two terms in the Missouri state penitentiary for burglary and had served nine months for grand larceny in Arkansas in 1912 but explained that he was a Pinkerton agent at the time and had pleaded guilty to protect his identity.[26]

The next day Hubert Vowells took the stand and testified he had accepted Krieger's alleged offer because he needed money to finance a planned bank robbery in Iowa. While admitting that he had never met Krieger, Vowells claimed Hall had told him what Krieger wanted them to do. He then said he and Benson had stolen seven sticks of dynamite from a quarry near Sand Springs and had hidden the explosives in a manure pile near their boarding house. The men planted the dynamite after midnight on October 29, 1917, he said, and then went to the home of an acquaintance, David Griffin. Vowells said he saw Hall talking to Krieger four days later and assumed it was over payment for the job, which he never received.[27]

But Moore substantially weakened Vowells's testimony during an eighteen-hour cross-examination. Under Moore's questioning, Vowells admitted he had never spoken to Krieger until both were in the Tulsa County jail in December 1918. Moore also produced letters written in the spring and summer of 1918 from Vowells to Hall, then serving time at Leavenworth for bank robbery, that never mentioned Krieger. The attorney noted Vowells first mentioned Krieger in a confession Vowells made in the Muskogee jail in October 1918 and made personally to Edgar Pew. Vowells admitted that he had confessed because Pew and other Carter Oil officials promised him they would use their influence to have him released and placed in the army. In addition, Moore demonstrated that another prisoner had received a similar pardon offer.[28]

The trial took an odd turn on October 28 when Moss asked to question Moore and Lowe on who had employed and paid them. Cole rejected the request, but the next day Moss demanded the two attorneys answer questions about possible jury tampering because E. M. Boyd and another Wobbly, Tom J. Jenkins, had talked to possible jurors in Bixby before the case. Moore said Boyd and Jenkins had acted without Krieger's knowledge, but admitted Boyd might have been overzealous. Testifying with the jury out, Boyd—still awaiting trial on the Wichita indictments—said he had come from Kansas City to investigate conditions for the union. He admitted having talked with four men on the venireman list but denied any bribery had occurred. Finding no evidence of tampering, Cole denied Moss's request.[29]

The defense relied on witnesses who testified that Krieger was working in Perry and Bristow for Mount Copper Boiler at the time the bombing occurred. W. G. Shipman, the foreman, and four other employees all said Krieger was with them in Perry unloading steel from railroad cars and helping build storage tanks. Walter Kerles, a Perry boardinghouse owner, said Krieger was in the hotel every night of that period, and Shipman testified he had roomed with Krieger from October 27 to November 14, 1917. Moore also produced a photo taken November 10 that showed Krieger as part of the work crew.[30]

But the most important witnesses were John Hall and George Benson, who were brought under guard from Leavenworth, where they were serving terms for robbing a post office. Although Lyons saw Hall as a Jesse James–style bandit, the forty-six-year-old Hall was a hardened criminal with a lengthy prison record for robbery and assault with intent to kill. As a result, Moore and Lowe initially expressed concerned over what Hall might say, but an interview with Hall persuaded them he was no informer. On the witness stand on November 3 Hall admitted stealing the dynamite but denied bombing the Pew home.

Instead he claimed he needed the dynamite to blow up some tree stumps, a story he maintained under cross-examination. More important, Hall also testified that Pew had visited him at Leavenworth and tried to bribe him to frame Krieger. "Pew said he would put me on easy street for the rest of my life," Hall said. He refused the offer, he said, because it was against his principles to frame anyone. Benson corroborated Hall's story, including Pew's offer, adding that he first met Krieger in jail in March 1918.[31]

On November 5, Krieger himself took the stand. He admitted that he was a Wobbly, that he was a job delegate and had taken part in a miners' strike in Humboldt, Arizona. But he denied the claims of Vowells, Hall, and Benson and said he had not spoken to Harper in jail because everyone knew the man was a "snitch." Asked about sabotage, Krieger said he believed only in peaceful strikes. After rebuttals—in which Pew denied offering bribes—the attorneys began their closing statements on November 8. Moss and assistant prosecutor L. G. Owens attacked the IWW as diabolical and perverted and claimed that only a person deranged by IWW doctrines could have committed the crime. In contrast, Moore, in a three-hour summation, told the jury that the facts showed Krieger was not in Tulsa at the time and that Vowells's testimony was false.[32]

On Monday morning, November 10, after forty-one hours of deliberation, the jurors told Cole they could not reach a verdict. They had taken three ballots: the first had been seven-to-five for conviction, the second was split evenly, and the final ballot was nine for conviction, three for acquittal. Lyons later learned one intransigent juror had opposed a conviction, in part because of a dispute with Standard Oil over a pipeline on his farm. Cole declared a mistrial and ordered Krieger returned to jail on a $2,500 bond. The *Tulsa Democrat*, through one of its reporters, paid the bail with Liberty Bonds collected from various sources. Why the *Democrat* put up the bond—other than perhaps to embarrass the rival *World*—remains unexplained.[33]

Krieger waited six months before his second trial began. In the interim, Moore located twelve new witnesses but lost his job when Bill Haywood and the GEB fired him after he missed a March 18, 1920, deadline to file a bill of exceptions in the Wichita case that could have freed the defendants. Moore was officially replaced by Harold O. Mulkes, although Moore stayed on to assist the new defense team.[34]

The new trial began on May 13, 1920, but jury selection problems delayed testimony until May 24. Edgar Pew opened the prosecution's case, testifying that he was certain his house was bombed because he had smelled nitroglycerine and had found fuse marks on the porch. Vowells reiterated his testimony but under Moore's cross-examination acknowledged that he had confessed in order to avoid more serious charges. In addition, Vowells admitted he had not met Krieger until well after the bombing and also revealed his total lack of knowledge about the Pew home. Although Vowells's admissions effectively wrecked the prosecution's case, they produced other witnesses to prove their theory of a bombing conspiracy, including a former Bisbee, Arizona, police chief whose testimony that Krieger had met John Hall during the copper miners' strike of July 1917 was intended to prove a conspiracy existed. But another prosecution witness testified Krieger and Hall had been in Woodward at that exact same time.[35]

After the prosecution's case unraveled, the defense again relied on Hall and Benson and Krieger's co-workers to establish Krieger's whereabouts both in July 1917 and on the date of the bombing. Hall again denied any conspiracy. Sheriff J. W. Green of Rogers County bolstered the claim by showing Hall was in the county jail from August 3 to August 17, when the alleged conspiracy was supposedly hatched, while a Claremore boardinghouse owner testified Hall could not have been in Bisbee because the felon had stayed at his boarding house from the middle of July to August 1917.[36]

The case went to the jury June 8 after summations. Prosecutors again referred to the IWW's alleged violent beliefs, while

Mulkes accused the prosecution of inventing a nonexistent conspiracy to frame Krieger. The jury sided with Krieger and, after two hours and three ballots, returned an acquittal. Charles Krieger was finally free after two and a half years and a cost to the IWW of $12,000. More important, the only attempt by authorities to use the legal system to prove the IWW had used violence had failed. The IWW leadership saw itself as vindicated and hoped to use Krieger's acquittal to counter the negative image purveyed by the government and the press. The General Defense Committee asked Krieger to make a speaking trip to raise funds to help pay off the remaining $4,000 of his court costs and to aid other jailed union members such as Jack Terrell, who was awaiting sentencing in Enid for criminal syndicalism.[37]

But the IWW's celebrations were premature. Although an important legal victory for the IWW, the Krieger case's real effects were limited and temporary. It certainly meant that in the future prosecutors would need substantive, physical evidence rather than perjured testimony to connect a Wobbly to a violent crime. But authorities had already moved on to another legal device to deal with the IWW, one not limited by the need for solid evidence.

If the Krieger case attempted to prove that mere IWW membership was prima facie evidence of labor violence, then the state's 1919 criminal syndicalism law accepted it as fact. The Oklahoma law itself was not unusual as state legislation to curb alleged labor violence had become commonplace by the war's end, especially as federal prosecutions of radicals declined. Oklahoma was one of eighteen states and territories, mostly western, that passed such laws. They derived in part from Australia's Unlawful Associations Act, which had specifically targeted the IWW and syndicalists in that nation. Minnesota, Idaho, and Montana approved the legislation first, but the idea spread throughout the West, especially in states that had witnessed bitter labor-employer conflict. All the criminal syndicalism laws were aimed at the IWW, based on the assumption that the Wobblies advocated violence, and nearly all the laws defined criminal syndicalism as any doctrine advocating crime, violence, sabotage, or other unlawful means to force industrial or political reform. Oklahoma defined sabotage as malicious, felonious, intentional, or unlawful damage to employers' property by employees or those inciting them to act and imposed a maximum ten-year jail sentence and a $5,000 fine.[38]

The attraction of such a law for Oklahoma was simple. It allowed authorities greater leverage in preventing labor activity in two of the state's major economic resources—wheat and oil—and it gave legal status to the image of the IWW as violence prone by making mere membership in the union an act constituting criminality. The cost of litigating such a law would add to the IWW's already financially tenuous position; and, those convicted under the law could not appeal to federal courts for civil rights protection as the Supreme Court at this time had refused to apply the Fourteenth Amendment to the states in order to enforce the Bill of Rights. The courts had also argued that victims whose rights had been abused should appeal to state authorities. As a result, federal courts refused to hear appeals from criminal syndicalism convictions.[39]

In early 1919, newly-elected Oklahoma governor Robertson urged the state legislature to approve a criminal syndicalism law for Oklahoma. That Robertson would want such a law is understandable. A business progressive and Democratic loyalist, he was a member of a law firm that represented Standard Oil of Indiana. In his inaugural address he had urged the establishment of loan programs to promote home purchases by workers as an antidote to radicalism and the IWW. Robertson's apparent fears of the IWW and labor radicalism probably explain his overreaction in sending the National Guard into Drumright during the telephone operators' strike and in declaring martial law during the coal strike of 1919.[40]

Robertson soon found an ally in state senator Luther Harrison. Editor of the *Wewoka Capitol-Democrat* and chair of the Seminole County Council of Defense during the war, Harrison also rode with a posse during the Green Corn Rebellion, a fact he used in his 1918 state senate race. A well-known racist and Red-baiter who urged the deportation of leftists, Harrison apparently conceived of an antiradical law as early as 1917. Although he later claimed the law was not aimed at either the IWW or the Socialists, that was patently untrue. In introducing the bill, he declared it would forever ban the IWW from Oklahoma—a prophecy that proved incorrect. But Harrison's timing was superb as the press spread tales of Red revolution in the wake of the Seattle general strike.[41]

The bill—based on Oregon's criminal syndicalism law—had little trouble gaining approval, although the Oklahoma State Federation of Labor successfully lobbied for an amendment to exempt boycotts, strikes, and picketing. The OSFL had hoped the amendment would effectively cripple the bill, but the legislature accepted the changes, perhaps because Congress had recently approved the Clayton Act, which gave labor limited rights to organize, strike, and picket. The legislature also approved the emergency clause so the law went into effect immediately, with no public referendum.[42]

Curiously, the state press—which had so vilified the IWW—gave little coverage to the debate on the bill and the amendment. Nor did the press offer many editorials regarding the new law. Governor Robertson thanked Harrison for introducing the law, although Robertson said it was not as strong as he would have preferred. But he was satisfied enough to declare, "It can harm no law-abiding citizen."[43]

The IWW knew whom the law was meant to harm. An editorial in the Wobblies' *One Big Union Monthly* declared such laws had the sole purpose of exterminating the IWW and silencing radical political agitation. Capitalism could no longer stand bad publicity, the writer argued, and to protect it required the abolition of freedom of the press, speech, and assembly. "It isn't syndicalism that is criminal, it is capitalism and instead of having anti-syndicali[s]t laws, we should have laws against criminal capitalism, if we should have any." As the IWW feared, the first target of the new law was a Wobbly—Jack Terrell, an AWIU job delegate arrested by Enid police in June 1919.⁴⁴

Terrell's arrest coincided with a renewed AWIU harvest campaign. Despite the financial burden of numerous federal and state prosecutions, the IWW planned a large-scale harvest effort for the summer of 1919. The union's leadership realized that large numbers of returning veterans were left jobless because the Wilson administration had failed to develop an adequate demobilization plan. The IWW saw the men both as a threat because their availability would inflate the labor supply and force down wages, and as potential members because of their frustration with the lack of government response. As one veteran told a Wobbly named "Doc" near Oklahoma City, "I thought that when I left Camp Funston I could step out and get a job at once, but here I am back at the old life again."⁴⁵

The AWIU expected farmers would take advantage of the surplus hands and the free employment bureaus would once again try to flood the fields with workers. Rumors had wages as high as $5.00 to $6.00 for a ten-hour day, but *New Solidarity* noted the pay in southern Oklahoma was only $2.50 a day, despite a bumper crop and high wheat prices. Union officials D. N. Simpson and Mat K. Fox urged Wobblies to position themselves near Enid, Ponca City, and Blackwell, Oklahoma, and Caldwell, Kansas, and to take general farm jobs while the wheat ripened. To ensure success, the AWIU placed a stationary delegate in Kansas City to coordinate recruiting and hoped to put a hundred job delegates in the field.⁴⁶

But the feared surplus failed to materialize and a relative labor shortage occurred, which should have benefited the AWIU. State and local authorities set up special police forces and armed gunmen to keep out the IWW, and wholesale arrests began the moment organizers appeared. Delegates were usually arrested for vagrancy, searched, and thrown into jail if they carried IWW cards, literature, or organizing material. Enid proved especially inhospitable, arresting over fifty men. Authorities singled out four job delegates (Terrell, George Aldridge,

Paul Vold, and W. H. Moran) and charged them with criminal syndicalism. The arrests led *One Big Union Monthly* to charge that Oklahoma had abandoned equality before the law and replaced it with vigilantism by the property-owning class in defiance of American principles.[47]

Of the four job delegates arrested, Jack Terrell was the most important. A thirty-five-year-old Ohio native with only a fifth-grade education, Terrell was the son of an English father and a French mother, both dead. Forced to work since age fourteen, he had joined the IWW in August 1917 while taking part in a threshing crew strike in Fargo, North Dakota. In 1919 he was elected to the AWIU General Organizing Committee and came to Oklahoma in May to coordinate harvest recruiting. Terrell was arrested June 15 outside an Enid boarding house, after the owner—who thought Terrell was a bootlegger—allowed a police officer named Marion C. Gross to search Terrell's rooms. During his warrantless search, Gross found a bag containing IWW material, which led authorities to charge Terrell with criminal syndicalism.[48]

At Terrell's preliminary hearing on July 10, 1919, attorney Fred Moore argued that Terrell's room and bags were illegally searched and therefore the charge should be dismissed. Garfield County Attorney Ernest J. Smith admitted the warrant was not issued until June 18, three days after Terrell's arrest. Despite Moore's arguments, Terrell and the other men were bound over for trial. The state's indictment, filed by Smith, rather ramblingly charged that Terrell did "unlawfully, willfully, and feloniously, organize and help to organize and did then and there become a member and did voluntarily assemble with a certain society, to wit, the IWW." The remainder of the indictment specifically tried to prove that mere IWW membership was advocacy of criminal syndicalism, which it also went to great lengths to define.[49]

Unlike Charles Krieger, Jack Terrell did not languish long in an Oklahoma jail. His trial began in mid-January 1920 with something reminiscent of Krieger's case: a lengthy jury selection. Moore and Smith argued over jurors, with Moore asking their opinions on the IWW, the Socialist Party, Standard Oil, and antitrust laws. Judge J. B. Cullison repeatedly rebuked Moore and accused him of trying to prevent the trial, but a jury was finally seated January 16.[50]

Local newspapers took the opportunity to attack Moore and the IWW. The *Enid Morning News* printed an editorial called "Stamp Out the Reds" reminiscent of the *Tulsa World* at its most vitriolic, while the *Enid Daily News* called Moore a professional IWW advocate who made the rounds for the union. The *Enid Daily Eagle* was the least shrill, noting that the lack of industrial unrest ensured an impartial trial and admitting Moore sought introduce testimony favorable to his client. But it alone of any state newspaper realized the trial's significance: "Being the first trial in the state under the espionage law passed by the legislature last spring, it will have statewide prominence and will be followed closely by all citizens who are anxious to arrive at the real truth as it relates to this much talked of organization."[51]

The real "truth," however, was not forthcoming in the rather short trial. Prosecutors relied exclusively on the material found in Terrell's bag, which Officer Gross admitted he had examined without a warrant. County Attorney Smith also admitted the warrant was not issued until after Terrell's arrest. In contrast, Moore challenged the seizure of Terrell's bag as a violation of both the Fourth Amendment protection against searches without warrants and the Fifth Amendment protection against self-incrimination. This now commonplace legal argument was quite novel for the time, although one reporter called Moore's tactic long-winded and rambling. It was also unsuccessful as Judge Cullison ruled Moore's objections incompetent, immaterial, and irrelevant.[52]

Cullison's ruling meant Moore's only recourse was to place Terrell on the witness stand. Terrell, who sat chewing gum through most of the trial, proved a surprisingly good witness. He denied teaching or advocating violence but said he did favor the use of force in the form of the general strike, which would force owners to surrender their property to the workers. To counter charges that all Wobblies were antiwar, Terrell testified he had registered for the draft at Sioux City, Iowa, but had not been called up, and when he tried to enlist at Chicago, he was rejected. The *Daily News* also expressed amazement at Terrell's articulateness despite his lack of formal education: "[I]t seemed as though the whole wisdom, ignorance, and forcefulness of the entire IWW organization sat in the witness box and bantered with the attorneys at will."[53]

But Terrell's testimony and demeanor failed to sway the jury, who found him guilty after deliberating only thirty minutes. Moore called the verdict unjust, blaming it on widespread prejudice against the IWW and the prosecution's attempts to tie all calls for economic change to violent revolution. On February 16, Judge Cullison sentenced Terrell to seven years in the state prison in McAlester. Moore immediately filed an appeal, and Terrell was released on bond.[54]

Terrell, however, never served a day of his sentence. His initial appeal was held on July 26, 1920, at which point Judge Cullison held the case over for a year on the condition that bondsmen be able to produce Terrell on sixty-days notice. In addition to effectively letting Terrell go free, the state declined to pursue charges against Vold, Moran, and Aldridge. Why this happened is unclear, but it may have stemmed from serious doubts over the legality of search or from concerns regarding the syndicalism law's constitutionality. Whatever the case, Oklahoma Attorney General George F. Short reportedly had admitted the conviction was in error, and the IWW leadership expected a dismissal.[55]

But no dismissal came. By the time the appeal finally went to the state Criminal Court of Appeals, Terrell had been convicted on criminal syndicalism charges in Los Angeles and was serving one to fourteen years in San Quentin, California. As a result, the court denied the appeal, and the law remained unchallenged and on the books. Three years later it resurfaced to plague the IWW as the union again tried to organize in Oklahoma.⁵⁶

As the Krieger and Terrell cases show, wartime political repression in Oklahoma did not end with the Armistice. If the war provided a means to wreck the state's strong Socialist and labor movements, then the peace continued such policies. What did change were the tactics. During the war probusiness forces relied on neovigilantism and the power of the federal government to suppress dissent. With the war's end the federal forces retreated slightly from direct confrontation with labor and radicalism, mostly because the government's wartime police powers had lapsed with the peace. While it could still deport radicals who were not American citizens, the federal government could rarely use this authority against the mostly native-born Wobblies.⁵⁷

States such as Oklahoma filled the void by enacting their own antisedition and criminal syndicalism statutes enforced by local courts. This, in effect, extended neovigilantism and gave it legal sanction. The state's business and political leaders—using the business-dominated county Councils of Defense, which acted as the de facto wartime state government—consolidated their hold on political power and opinion and sought to legitimize it through legislation and police power. Such action was a natural progression as state and local officials had traditionally been the agents of suppression of dissent, usually through the "creative" interpretation of state laws and city ordinances. Criminal syndicalism laws simply gave the state the same powers previously used by the national government.⁵⁸

Oklahoma officials may have also feared a revival of radicalism, which they believed threatened the investment capital needed for development. For while most of the states that adopted criminal syndicalism laws had long histories of labor-business conflict, few had experienced the level of organized resistance to incorporation that had existed in Oklahoma. Even with the influence of the Councils of Defense, support for the war proved lukewarm in the state as some counties failed by wide margins to meet Liberty Bond quotas and National Guard enlistments stayed low. Although the war wrecked the state Socialist Party and destroyed the Working Class Union, many Socialists after the war continued their activities even without a formal party, and remnants of the WCU continued to dynamite dipping vats. Further, there was a renewed appearance of the IWW. The Democratic Party itself remained split between the incorporationist business progressives and their agrarian, populist opponents. That fact was demonstrated first by the 1918 reelection of Senator Thomas Gore, an opponent of the war, mostly on the strength of rural voters although the Councils of Defense blamed it on "pro-German propaganda."[59]

The use of courts and new laws was more than just an attempt to stem radicalism. Having now gained the upper hand politically and economically, Oklahoma's business progressives sought to impose their version of politically acceptable discourse, exemplified by Victor Harlow's comment regarding draft resistance in 1917: "[F]ree speech is often mistaken for license to say anything." As in other states, Oklahoma authorities learned that the war's martial spirit readily adapted itself to new purposes, easily shifting from the "dreaded Hun" to the "dangerous Bolshevik" or the IWW. In fact, it may have proven easier in Oklahoma, for the alleged "pro-German" and "Bolshevik" propaganda was homegrown, and it took little effort to shift the emphasis from censoring one to restricting the other. The criminal syndicalism law, therefore, merely extended wartime censorship and outlawed any form of speech or protest deemed to threaten the prevailing economic interests.[60]

The trials of Charles Krieger and Jack Terrell, then, represented legalistic ploys to define the IWW and radicalism as beyond the pale of American society, indeed as "un-American," as new usage had it. In Krieger's case, the prosecution's (or, rather, Carter Oil's) zealous pursuit of a conviction at any cost—including purchased, perjured testimony—ran afoul of a conscientious judge and the legal requirement for evidence proving Krieger had conspired to bomb the Pew home. The facts of the case mitigated against Carter Oil. For Terrell, the problem was more sinister. Criminal syndicalism defined mere membership in the IWW and possession of union literature as criminal. This relieved the prosecution of the burden of proving that any other crime had occurred. Terrell's membership card, his bag of literature, and his belief in IWW ideology were all the evidence needed.

The results were a mixed blessing for the IWW. Krieger's acquittal meant authorities in the future had to prove that a Wobbly had committed a real crime, but Terrell's conviction meant union members had to walk a fine line in what they said or did or face imprisonment for their beliefs. Regardless of the threat, though, the IWW elected to continue its activities in Oklahoma.

7

A Brief Renaissance
Harvest Stiffs and Roughnecks in the Postwar Period

WITH the end of the war, the IWW tried to reestablish itself as a functioning union. It was aided by economic hardship and increased labor unrest following the Armistice. Agricultural prices collapsed as the government stopped subsidies necessitated by the war effort. Industry executives reneged on agreements negotiated by the War Labor Board, slashed wages, and refused to recognize unions. Mechanization, spurred by wartime military spending, continued to increase, helping create growing levels of technological unemployment. Demand fell in the oil industry, despite growing numbers of automobiles and the conversion of railroads from coal to fuel oil. In addition, the Wilson administration's lack of an adequate demobilization plan meant many veterans returned to home to find no jobs waiting. But in such chaos was opportunity, and the IWW seized the chance, especially in the wheat and oil fields of Oklahoma. In the resulting brief renaissance the union attained some of the highest membership totals in its history.[1]

Many historians of the IWW, however, have dismissed or ignored the union's growth in the postwar period. Several assumed that federal repression had left the union a rump organization torn by petty squabbles, its large membership roles the result of old members paying dues out of nostalgia rather than from new recruits. This image of the postwar IWW is less that of a union and more that of a radical fellowship. Nothing could be further from the truth.[2]

Certainly the federal prosecutions essentially decapitated the union leadership, but the IWW still had a core of veteran members and organizers such as Tom Doyle, Joe Fisher, E. W. Latchem, and W. R. Parker, the latter a longtime organizer in the Oklahoma oil fields. Perhaps less skilled than the older leadership, these men were far from incompetent, and many were encouraged by the growing postwar strike wave. Because the IWW was financially strapped and politically hounded, the leaders had their work cut out for them. But to rebuild the IWW, they turned to the industrial unions that had served the IWW so well—the AWIU and its affiliate, the OWIU.[3]

With the 1919 harvest came the first change that would revitalize the IWW. The 1918 harvest suffered both from a war-caused labor shortage and from the use of inexperienced volunteers to fill the gaps. Making matters worse, Oklahoma farmers had expected high wheat prices and had overplanted. To solve the problem, state officials relied on youngsters recruited into a Boys' Working Reserve force based on military lines and on "twilight crews" composed of businessmen who volunteered after hours. Some farmers even hired black harvesters for the first time. But while wages ran from four dollars a day for barge men to six dollars for stackers, train fares of three cents a mile cut deeply into workers' earnings despite official requests for half-price fares for harvesters. None of this satisfied the farmers, who found it required two inexperienced volunteers at twice the wages to do the work of a single, veteran migratory. They also knew that wartime voluntarism was unlikely to continue.[4]

The AWIU believed this would work to their advantage in 1919, especially because in 1918 it failed to initiate any new members in Oklahoma, Kansas, or Nebraska and had only about two dozen delegates in the field. The union expected 1919 would bring exaggerated requests for harvest hands and promises of five to six dollars for a ten-hour day. The result would be fields flooded with men, many of them returning veterans, which would lower wages. The AWIU believed dissatisfaction with the low pay would grow, as would the level of recruits. But success depended on having a large number of organizers, so the union's new secretary-treasurer, Mat Fox, gave job delegate credentials to anyone who had been in the union for six months. As an editorial in *Solidarity* noted, "If the Farmer had a 'bumper crop,' let us get in and organize the agricultural workers for a 'bumper membership' and 'bumper wages.' "⁵

But the AWIU immediately encountered resistance as wartime antiradicalism continued unabated. Even before the harvest began police had arrested nine Wobblies at Wynona and Pawhuska and seized the IWW literature they carried, while Bartlesville authorities raided a hobo jungle in February after IWW stickerettes appeared in that community. Oil workers were arrested and jailed, and their cards and supplies were confiscated in Drumright and Cleveland, Oklahoma. Fifty men, including Jack Terrell, were jailed in Enid after the harvest began, and police arrested Wobblies in Medford and other Wheat Belt towns. Kansas alone jailed at least one hundred job delegates. At the harvest's end, the AWIU estimated a thousand Wobblies languished behind bars.⁶

Despite the AWIU's expectations, the high wages held rather than fell, even with returning soldiers swelling the harvesters' ranks. But the AWIU did add four thousand new members and saw its cash reserve grow from $3,269 at the start of the year to $11,927 by November 1. This enabled the AWIU to donate to several IWW newspapers as well as help strikers in Brooklyn. The 1919 campaign proved a minor success at best and hardly offset other membership losses, but it still encouraged the IWW. In fact, the resistance from harvest town citizens may have seemed like old times to the veterans and contributed to a renewed enthusiasm for the union.⁷

As a result, the AWIU redoubled its harvest efforts in 1920. But conditions had changed little from 1919. Once again the fields were filled with workers after the state labor commission called for 10,000 harvest hands to bring in three million acres of wheat. The free employment bureaus provided 8,765 harvesters, and wages remained high, with men at Enid receiving

$6.00 to $7.00 a day, although $4.50 to $5.00 was more common. But with wages high and the labor supply more than adequate, harvest stiffs had little use for union agitation, although some areas saw Wobbly-led job actions and work slowdowns to protest poor conditions. Local authorities also continued to arrest Wobblies. George Aldridge, who was charged with but avoided conviction for criminal syndicalism in Enid in 1919, was arrested in Medford with three other IWW members and accused of murdering a railroad policeman. The charge proved false, and he was released in August, long after he might have been able to help with recruiting.[8]

In response to the poor performance, the AWIU adopted several measures to strengthen the union's effectiveness and financial status. Job delegates would now receive a one-dollar commission on initiation fees, up from fifty cents, and stationary delegates were given the power to issue credentials and supplies, an authority previously limited to the General Organizing Committee. The union also decided to provide monetary assistance only to the IWW as a whole and to the defense committee rather than to individual jailed members. It also froze officers' wages and accepted a symbolic change with a redesignation from its old no. 400 to no. 110, a number more consistent with IWW branch designation practices.[9]

The changes came at an appropriate time. By the end of 1920 an agricultural depression had set in, with the price for Oklahoma wheat dropping from a wartime high of $2.54 a bushel to a low of 85¢ a bushel by 1921. Farmers complicated matters by overplanting to compensate for the reduced prices. When heavy rains set in before the harvest, state labor officials realized Oklahoma faced a labor shortage. Wages fell to $3.00 to $4.00 a day, barely above prewar levels. Although the free employment bureaus recruited 11,296 men, the total proved inadequate, and the shortfall aided the AWIU.[10]

The IWW entered the fields in greater numbers than in anytime since 1917. The job delegates changed their strategy and focused less on the larger towns and more on smaller communities that suffered most from a lack of hands. The main goal was permanent establishment of the ten-hour day, but delegates at a meeting in Enid wanted to push for a $7.00 daily wage. These goals were overly optimistic as pay started at $3.00 to $3.50 at best and working hours far exceeded ten. The Wobblies also spent much of their energies dealing with local police rather than organizing. They managed to free ten jailed workers at Enid, although four delegates remained in jail until August. A delegate in Lamont, in Grant County, also noted the union needed to start recruiting threshing crews as well as harvesters if it expected to win the ten-hour day.[11]

But AWIU organizers were exhilarated even by what little success they had achieved. Wages in some regions had increased by fifty to seventy-five cents, and the AWIU began comparing the campaign to 1916, the agricultural workers' union's best year. Between July and August delegates had distributed 160,000 leaflets, 40,000 pamphlets, and thousands of newspapers. By August the AWIU had recruited 5,000 new members.[12]

The 1922 campaign in Oklahoma got off to a slow and rocky start, compounded by a poor harvest and increased local resistance. Perhaps because of the price collapse, the crop was short that summer, requiring fewer hands. At the start of the harvest large numbers of unemployed men remained in Oklahoma, with five hundred in Enid alone. Another seven hundred had already left to find work in Kansas. The AWIU, as result, lowered its goals to a more realistic hope of raising wages to five dollars a day. *Industrial Solidarity* called for the drive to begin in southern Oklahoma rather than Enid in order to build momentum and to take advantage of the large numbers of unemployed. "Most of the workers making the harvest have had a good dose of capitalist democracy and freedom (to starve) last winter, so they will be more amenable to the organization's propaganda than in former years," the newspaper declared.[13]

Industrial Solidarity's predictions seemed borne out by the attendance at meetings in jungle camps in Enid and Alva in early June. Two Enid meetings, on June 11 and 16, attracted nearly two hundred harvesters, about half nonunion members, who agreed to work for five dollars for a ten-hour day. The plan called for harvesters to take available jobs, work for one day, then begin their strike between 9:00 A.M. and 10:00 A.M. the next day until their demands were met. The strikes were designed to cause the greatest inconvenience for farmers who could lose an entire day's harvest if they failed to raise wages.[14]

Local authorities learned of the plans before they could be implemented. In early June, Enid police arrested one job delegate on vagrancy charges, but released him after a hundred or so harvest hands crowded into the local courthouse. The show of solidarity briefly deterred arrests, but police arrested fifty-seven men, mostly IWW members, on June 11. Six other Wobblies were also detained in Alva and Waynoka on June 9. Those six men were released without incident, but the Enid prisoners faced vagrancy charges. They were arrested when the county sheriff, deputies, and railroad police raided the IWW jungle near the Rock Island tracks after a 3:00 P.M. meeting had ended. The sheriff asked which men were organizers. When the IWW members replied, "We're all organizers," they were jailed. Twenty-two were placed in the county jail, with the remainder in the overcrowded city facility.[15]

The men were arraigned on June 12. Six men who were not Wobblies were released, while the others—including two men arrested later—refused to plead until they spoke with an attorney. What resulted was an absurd situation in which Enid authorities, with the jails filled, could arrest no one else and had to tap into a special fund to feed the prisoners, who were daily marched to the county courthouse for meals. This forced the assistant county attorney to order the remaining jobless men to leave town or be forcibly rounded up and driven across the county line, an action that might have been difficult to implement, given the large numbers of unemployed.[16]

The arrival of attorneys Harold Mulkes and John J. Carney on June 15 resolved the situation. The charges were dismissed after the men who had been arrested agreed to leave town, and local officials admitted the arrests were improper because the men had wanted jobs and had registered at the employment bureau. Although the organizers had been forced from town, their efforts were praised by the editors of *Industrial Solidarity*, who saw the size of the camp meetings, the arrests, and the vagrancy dismissals as a symbolic victory for labor unity. Too enthusiastically, they declared, "Enid is a milestone in the history of labor. It is not our destination; merely an incident, marking where we camped." The newspaper urged organizers to take up the spirit of Enid and sign up more members. "Delegates and members, we have only misery to lose. We have decency, comfort, a human living to win."[17]

The enthusiasm was tempered, however, by the murder of an AWIU job delegate on June 16 in Cherokee, in Alfalfa County. The incident occurred in a city park when seven Wobblies confronted three men, not union members, whom they believed had "finked" to Garfield County and Enid authorities, leading to the June 11 arrests. A fight ensued in which one man pulled a gun and shot Wobbly Paul Bedmarcek—a Polish-born organizer and a war veteran—in the heart. The Wobblies claimed the gunman, a non-Wobbly named Tom Pryor, was "coked up" and had started the incident, but Pryor claimed Bedmarcek had a knife. Although no knife was produced, a coroner's jury exonerated Pryor. The AWIU members, though, were herded onto a train by the sheriff and forced to leave town.[18]

The AWIU remained undiscouraged despite the evidence of growing opposition in Oklahoma. On the contrary, events spurred the organization, which—if union papers are to be believed—used "Remember Enid!" as a battle cry throughout the summer of 1922. Thus motivated, the AWIU acquired 14,459 new members and raised a surprising $135,055. The Oklahoma incidents may have also encouraged the AWIU to hold its spring conference in Oklahoma City in May 1923, although the presence of the headquarters of the Oil Workers Industrial Union in the city certainly aided in its selection. In addition, the Wobblies probably believed the 1922 election of a prolabor state governor—John C. Walton, supported by the Farmer-Labor Reconstruction League—augured well for their reception. Whatever the case, the 1923 convention would mark the start of the AWIU's best campaign in the postwar period.[19]

The conference convened May 20 at the AWIU's offices in a red brick building on West Reno Street. Fifty-six union members from Oklahoma, Kansas, and southern Nebraska and six visitors attended the convention, which opened on May 20. Tom Wallace, AWIU secretary-treasurer, told attendees that at that time the union had only $3,156 left over from 1922, with another $23,948 on hand. While the low amount suggests a possible economic crisis for the AWIU, it actually indicates how dependent the IWW as a whole had become on the agricultural workers. Nearly $85,000 of the money the AWIU had raised in 1922 went to support strikes in Oregon and California; and harvesters had also provided defense funds for Wobblies in Centralia, Washington, and made a large donation to the IWW's Work People's College in Duluth, Minnesota.[20]

During the conference the delegates approved a proposal for a general strike of harvest hands if Kansas authorities tried to disrupt IWW activities or utilized that state's hated industrial court to prosecute union members. They also agreed to cooperate in any general strike called to release class-war prisoners. Attendees again endorsed the ten-hour day for harvest hands and approved a 50¢ commission for each new member—a decrease from 1920—and created voluntary traveling delegates who received $10 a week. In addition, the delegates elected to hire speakers at $28 a week to recruit in North Dakota and to spend $3,000 to $5,000 for organizing campaigns in California. They also planned to begin recruiting efforts in the Corn Belt states east of the Mississippi.[21]

Such actions reveal a high degree of optimism among the AWIU leadership. But other actions indicate the growing tension between the AWIU and other industrial unions. The delegates rejected proposals for special $1 dues stamps to cover legal defense costs and support the union's newspapers. Instead, the members approved only a special stamp to defray the AWIU's $13,150 portion of the IWW's bail and bond debt. They also rejected proposals for compulsory four-dollar assessments to aid organizing and refused to supply $3,000 to aid striking members of the lumberjacks' and Marine Transport Workers unions.[22]

Besides suggesting that the AWIU did not wish to support the organizing costs of other industrial unions, the actions of the Oklahoma City convention delegates indicated that the AWIU was unwilling to pay for other unions' legal costs. Significantly, the only defense assessment it supported was the one most important to it—bail for jailed members. These resolutions hardly endeared the AWIU to other industrial unions, which also suffered from state and federal repression and which lacked and resented the agricultural workers' resources. Such longstanding resentments soon contributed to a serious split in the IWW.

But that was to come. The convention adjourned on May 24, with the thirty-nine delegates there singing the labor song "Hold the Fort." The delegates' mood was undoubtedly buoyant, helped by the relatively calm reception Oklahoma City gave the union. Just two Wobblies were arrested on vagrancy charges and they were fined only twenty dollars. Still, Mayor O. A. Cargill said police would now investigate IWW activity in the city, and the *Daily Oklahoman* saw sinister designs behind the convention and asked what had happened to the state's criminal syndicalism law.[23]

With the convention over, the AWIU urged members to head to Alva and Enid, hoping to take advantage of a labor shortage and nearly 3.4 million acres of wheat. The Oklahoma Labor Commission said the state needed 28,000 harvesters—7,500 from out of state—but wages had dropped to prewar levels of $2.50 to $3.00 a day. Officials worried the railroads would worsen the shortage by denying rides to migratories, as they had in 1922. Given such factors, the AWIU expected that Oklahoma wheat farmers, fearing they would not obtain adequate help, would be more timid.[24]

Local officials, however, were far from timid. Kay County officials announced an all-out effort to stop IWW agitation after deputies arrested three oil workers' union organizers in Tonkawa and Ponca City, while in Garfield County, Sheriff Ora Lincoln and County Attorney V. P. Crowe declared "war" on the Wobblies, arresting four organizers and stationing a deputy at the post office to watch for men receiving mail from the IWW. Five more Wobblies were arrested that way. Soon thirty Wobblies and sympathizers were in the county jail; and *Industrial Solidarity* noted that county officials were far more worried about the IWW than about the bootleggers and other criminals preying on the migratories. It suggested that the American Legion and the Ku Klux Klan—the latter a growing force in Oklahoma—were dictating the arrests. "The cardinal principles of constitutional Americanism, the guarantees of American freedom that our newspaper friends talk about, have long since found the scrap pile," an editorial in the paper declared.[25]

Authorities may have feared the Wobblies, but the labor shortage benefited the AWIU. Wages rose to $4.50 a day as desperate farmers sought help. In Enid, as members flooded into the city and swelled the jail population, the union used a tactic from the free-speech fights: the jailed union members began a nightly sing-along, joined by other Wobblies on the courthouse square. As additional arrests only further crowded the jail, the sheriff faced a crisis. It was resolved on June 16 when he met with the IWW's attorney, John J. Carney, and a three-member committee. Carney told Lincoln that unless the men were released, no workers would shock the wheat. The sheriff, seeing the need for the hands, released the men unconditionally and returned their union cards and supplies. When one delegate had his supplies returned, he quipped, "Victory feels good. Like Chesterfield cigarettes, it satisfies."[26]

The AWIU considered the Enid incident a clear victory and took credit for the wage increases, one of the few increases after 1920. Although the union probably gained few new members in Oklahoma, the AWIU used the incident to increase enthusiasm among its members, among whom it likely renewed memories of old times. It certainly aided recruitment in other states, with new memberships reaching a postwar high of 15,217. At the union's fall convention in Fargo, North Dakota, the delegates approved plans for a "four-in-one" drive for 1924 that would begin in Oklahoma and recruit not just harvest workers but railroad, construction, and lumber workers as well. "Nothing could stop it," one delegate exclaimed.[27]

But there were also warning signs. By the fall, AWIU expenses totaled more than $193,000, while receipts totaled only $174,000. Including cash on hand, the union had only $4,871, hardly enough to fund such an ambitious campaign. In addition, the union had problems in the northern part of the 1923 campaign because of a poor wheat yield and competition from financially-strapped farmers from Oklahoma, Texas, and Kansas who rode north to earn extra money. It was a portent of things to come. But, in the wake of Enid and the early successes on the southern plains, the AWIU saw only a bright future ahead.[28]

In contrast to the AWIU, the Oil Workers Industrial Union recovered more slowly after the war. Still tied to the AWIU, it had fewer financial resources on which to draw. Perceived as more of a direct threat to the war effort, the OWIU suffered more from federal prosecutions than did any other IWW branch. The Wichita indictments left the union's leadership and its best organizers, such as Frank Gallagher, behind bars long after the war's end. Complicating matters, oil work declined as the wartime demand for petroleum dropped off quickly. Although a renewed oil boom later made up the difference, the market became oversaturated, causing prices to collapse and eliminating jobs.[29]

Nonetheless, the IWW still saw the oil fields as a major area for organization. IWW publications such as *Industrial Pioneer* and *Industrial Solidarity* devoted entire issues to oil workers; *Industrial Solidarity* also ran a series of articles covering all aspects of labor in the industry. One article noted that railroads operating in

Oklahoma were shifting from coal to fuel oil, causing the closure of mines at Lehigh and Hartshorne. Such changes suggested to the IWW that an energy trust would soon dominate the nation, just as union theorists had expected, and the Wobblies should prepare to take over when the system inevitably collapsed. Writing in *Industrial Pioneer*, Albert Barr—recently released from Leavenworth—told oil workers that the task would take "hard thinking, hard working and hard fighting."[30]

The work proved harder than the OWIU anticipated. In 1919, the oil workers union had little success in the Midcontinent fields as the postwar economy left as many as five men seeking each job. Authorities continually harassed Wobblies, such as three delegates arrested in Muskogee who narrowly avoided being charged with plotting to overthrow the federal government. Overall, job delegates in the Oklahoma oil fields faced arrest more often than harvest organizers did. While this is partly explained by oil workers' remaining on the job longer than migratories, sometimes police in places such as Bartlesville, Drumright, and Muskogee seemed to know exactly when Wobblies came to town. A delegate named "Doc" reported that he was arrested the moment he arrived in Cleveland, in Osage County. "I was at once confronted by the limb of the law, taken to the hoosegow, quizzed by Hizzoner and told that I was a dangerous person to be at large," he told *New Solidarity*. After twenty-five hours in jail, he was released, minus his supplies and union card.[31]

The economic situation worsened between September 1920 and September 1921, when 10,500 Oklahoma petroleum workers—about 32 percent of the workforce—lost their jobs. Those workers who remained saw their wages drop from $5.00 a day for eight hours to between $3.50 and $4.00 for ten- to twelve-hour shifts. Furthermore, Standard Oil of New Jersey had begun an aggressive campaign to establish "company unions." The idea spread to the Midcontinent, where Standard's subsidiary Carter Oil and companies such as Sinclair, Phillips, and Indian Territory Illuminating Oil adopted it. The possibility of organization seemed futile even to R. E. Evans, president of the AFL's International Association of Oil Field, Gas Well, and Refinery Workers. In 1920 he estimated only 15 percent of the Midcontinent workers were unionized, and many Oklahoma locals had collapsed from lack of dues revenue.[32]

But the IWW still believed the oil industry could be organized. They gained their chance when the California branch of the AFL oil workers' union fought a series of strikes in 1920 and 1921 to hold onto its wartime gains. Although the strikes failed and the union fell from 24,800 members to 6,100, the strikers' actions impressed the Wobblies. The California workers adopted such tactics as flying squads, flexible picketing and a militarized picket line, and their own system of armed guards. The strikes also utilized radios and a blinker system using car headlights or heliographs to send messages and move reinforcements where needed. Wobbly Nick Wells admiringly compared the tactics to those of the IWW in the Lawrence and Paterson strikes of 1912 and 1913.[33]

Another IWW writer, William Dimmit, argued that the AFL union's lack of activity had disgusted its membership. It had done nothing to improve conditions in the Midcontinent and proved unable to help the California workers. Only the IWW had the means to help drillers and pipeliners end the twelve-hour "tower" and the ten-hour shift. Editorials in *Solidarity* also pointed out that many Association members were former Wobblies who joined the AFL only after the wartime suppression of the OWIU. "The oil worker needs organization, and needs it badly. He is, in his way, a fighter, but he [has] always fought in an unorganized manner and, consequently, has gained nothing."[34]

As Wobblies saw it, the source of that organization was, of course, the IWW itself, and by 1920 the union had already made moves to strengthen the OWIU. A major change occurred in August when the GEB renamed and renumbered the OWIU, calling it the Oil, Gas, and Petroleum Workers Industrial

Union no. 230. The leadership also removed the union from the control of the AWIU and placed it under the mining department. The AWIU formally accepted the separation at its fall convention in New Rockford, North Dakota, and donated $1,500 to help the oil workers. The new union, however, remained dependent on the agricultural workers and did not really come into its own until the fall of 1921, when delegates to the AWIU's fall convention in Omaha voted for an additional $2,000 loan and approved establishing an OWIU district headquarters in Oklahoma City. The delegates appointed Henry Bradley temporary secretary-treasurer at a salary of $4 a day, and named a five-member General Organizing Committee, who would each receive $10 a week. The new OWIU set its first meeting for November 1, 1921 at the new headquarters in Oklahoma City's Oil Exchange Building.[35]

Almost immediately, the new union found itself taking part in a strike. Oddly enough, it was not an oil field strike but a walkout by Oklahoma City packinghouse workers against the Wilson and Morris plants in December 1921. The strike was part of a nationwide work stoppage called by the AFL's Amalgamated Meat Cutters and Butcher Workmen to protest arbitrary wage cuts and attempts to establish company unions by the major packing firms. The strike quickly turned violent, with several battles between strikers and strikebreakers in mid- and late December.[36]

OWIU members joined the strikers on the picket lines, and job delegates reportedly recruited a few workers for the IWW's Foodstuff Workers Industrial Union. Overall, though, the Wobblies simply cooperated with the strikers and freely and willingly suffered the same abuses as the AFL members. For example, on Christmas Day, 1921, company guards attacked three Wobblies on the picket line. The three men—members of the World War Veterans who had been gassed in the war—were beaten unconscious with revolver butts. Strikers vowed retaliation, which led the city's open shop supporters to demand that Governor Robertson send in the National Guard. The intervention was averted only when state Democratic politicians told the governor it could erode labor's support for the party, which had suffered major electoral defeats in 1920.[37]

The metropolitan press continually attacked the IWW, as did City Attorney Forrest Hughes, who threatened to arrest all the Wobblies in town. When a black strikebreaker was lynched in January 1922, the *Oklahoma News* claimed—without proof—that the IWW had committed the act. The *News* failed to retract the story when the guilty parties were revealed to be a racially mixed group of eight strikers, none of them Wobblies, who apparently had a personal grievance with the victim. Although the lynching turned sympathy away from the workers and contributed to the strike's collapse, the newspaper attacks helped the IWW. "It feeds and waxes fat on the lies and distortion of the capitalistic press," observed Oscar Ameringer, the editor of the Socialist *Oklahoma Leader*.[38]

Despite the packinghouse strike, the OWIU did not neglect the oil fields. At the November 1921 Oklahoma City meeting, delegates adopted a wide-ranging plan to recruit in Oklahoma, Arkansas, Texas, and Louisiana. They approved a new tactic for organizing in which active members and job delegates would organize section by section, supported both the OWIU and main headquarters. The tactic worked well enough that the OWIU was self-sustaining by March 1922. This allowed the union to pay back the $2,000 loan from the AWIU and to publish two organizing pamphlets of its own. The four-page handbills—"Fellow Workers" and "Oil Workers!"—gave information on the OWIU and urged AFL members to join and work for job control. The pamphlets became commonplace in the Oklahoma oil fields.[39]

Much of the OWIU's initial success came from the weakness and virtual collapse of the AFL oil workers' union. The Wobblies' militancy and their use of short, swift strikes to gain their demands gave the IWW an image of effectiveness that the AFL's International Association lacked. Certainly the terrible working conditions had little changed since the war's end, and wages had hardly improved. An Oklahoma socialist, quoted in *Industrial Solidarity*, said the growth of the OWIU seemed steady and not at all what he considered "mushroom" or "revival-meeting" growth. He added that he thought the union attracted a "sober thinking element."⁴⁰

The OWIU also worked on its organizational strategy to take advantage of this. At an April 1922 conference in Oklahoma City, the OWIU's general organizing committee (GOC) divided the oil fields into five districts to improve recruiting, with Oklahoma and southern Kansas making up District One. GOC member Nick Schwartz was named delegate for District One, and the GOC elected to strengthen the leadership of the union by requiring officers to have served as job delegates and to have carried IWW literature for six months. While the union was now self-sustaining, it borrowed $985 from the AWIU to fund recruiting. The new strategy seems to have worked, for by July 1922 the OWIU claimed its membership was growing in all areas but Arkansas and Louisiana.⁴¹

While the OWIU still worked mostly with pipeline workers, the union began attracting a new element—the tank builders. This was a major change in membership. Generally considered skilled workers, the tank builders, or "setters," needed to know carpentry, construction, and metalworking because most storage tanks were then built with a wooden framework over which sheets of wood or metal were riveted or bolted, with the seams sealed with caulk. The work was dangerous, as "tankies" working on the structures ran the risks of falling from the scaffolding, being blown off by the wind, or being crushed by falling construction material. One veteran tank builder said a tankie could be recognized by the scars on his head and body and by the broken veins in his arms and face caused by heavy lifting.⁴²

Other oil workers perceived the tank builders as clannish and rowdy, hardly suitable for unionization; but the men worked in teams, and this served to increase solidarity among them. In fact, before the First World War most tank builders had belonged to the AFL's boilermakers' union, which had job control until a 1914-15 strike, when the oil companies refused to renew a union contract and imposed a piecework system. The tank builders' position was further eroded after the war when the oil corporations farmed out the construction work to subcontractors. Those contractors extended the use of piecework because it reduced labor costs, created a pool of surplus labor for potential strikebreakers, and gave the contractors more control over the job.[43]

Theoretically, piecework offered workers the opportunity to earn several times the wages of "bit," or hourly, work, but the rates actually dropped as more work was done. A crew in Enid learned this when they tried to see how much they could earn. After the first day, the employer had cut the rates. The men quit but were easily replaced by unemployed workers willing to work for less. One OWIU member noted that on one Oklahoma job a tank builder on piece-rates caulked 1,400 feet of an 80,000-barrel tank in a day, which earned him $28, or four times the hourly rate. But to earn that amount, the writer said, the caulker had done work equivalent to that of twelve or thirteen men. Piece-rates therefore cheated workers of wages and denied other men jobs. More important, as the Wobblies and the tank builders both knew, this form of so-called scientific management increased the power of the contractors. The OWIU called for a return to the bit system, which they also argued would also provide jobs for unemployed men willing to work and would increase job safety.[44]

The appeals based on job control and employment undoubtedly caused many tank builders to join the IWW. In addition, as skilled workers, the tank builders, when organized, were in a better position to enforce their demands than were other oil workers. They demonstrated this by winning two strikes in the late autumn of 1922. In DeNoya, tank builders struck a Skelly Oil project after six men were dismissed for being Wobblies. When the foreman refused to rehire the Wobblies, the other fifty or so men stopped work. According to a report in *Industrial*

Solidarity, the strike lasted ten minutes. Because a lack of good tank builders in the field meant the contractor could not replace the entire crew, the foreman reluctantly reinstated the six. A November 15 strike by seventy-five men in Hominy lasted three days. In that walkout the riveters demanded higher piece-rates for using larger rivets on tank bottoms and demanded extra pay for moving the metal tank sides—called "door sheets"—to the site. As in the first strike, the employer gave in to the demands. A second strike in DeNoya occurred in late December after a scaffolding on a tank collapsed, killing two men and seriously injuring two others. When the company refused to install the guardrails required by state law, OWIU delegates signed up the entire crew.[45]

By the fall of 1922 the OWIU was well on its way to becoming an important part of the IWW. At its first annual convention, held in Oklahoma City from October 16 to 19, 1922, the OWIU could boast thirty-four new delegates and $4,000 worth of supplies sent to Casper, Wyoming, as well as twenty-four job delegates in California. But District One, comprising Oklahoma and Kansas, was the most successful area, with forty job delegates and an income of $4,257 from supplies and dues stamps. Overall the union had 900 members and 161 delegates, with an income of $20,000—small numbers compared to the AFL oil workers' union's total of 6,100 members in 1922. But the AFL's strength was dwindling, its membership dropping to only 700 four years later. The Wobblies' strength was in the Midcontinent and appeared to be increasing. Compared to other branches of the IWW, the oil workers were also a sizable union that could exert increasing influence on the IWW as a whole. This led Wobbly Edward E. Anderson in a poetic moment to claim that the OWIU was "[a] day nearer that day when the Black Gold as it gushes forth from the earth shall flow to enrich the lives of the workers instead of the coffers of King Oil."[46]

Anderson, it turned out, was wildly optimistic. The OWIU faced serious difficulties. Although self-sustaining, it was far from financially stable. Delegates and members faced vigilante attacks from the Ku Klux Klan, and authorities revived the dormant criminal syndicalism law. Internal strife within the IWW leadership would also weaken the OWIU. Additionally, both the oil workers and the agricultural workers began meeting technological changes they could barely counter. The postwar revival would prove short-lived.

Still, the revival contradicts most views of the IWW in this period. Many historians have believed that wartime repression deprived the organization of its strong leadership, and that the IWW fell into the hands of less-skilled men, marking a complete break with the past. Others have argued that the various federal and state prosecutions transformed the IWW into a legal defense organization like the American Civil Liberties Union (ACLU) and that it never recovered as a union.[47]

But the Oklahoma experience indicates just the opposite. While the best-known leaders of the IWW may have been jailed, those men who replaced them were far from novices. Most were veterans of earlier IWW campaigns and several, like E. W. Latchem, had more experience with the daily operations and tribulations of the IWW than did leaders like Haywood or Chaplin. Perhaps lacking the revolutionary zeal of the old leaders, these new officials were no less committed to IWW goals. On the whole, they were more concerned with goal-oriented issues than they were with the revolution.

Indeed, the rebirth of the AWIU and OWIU after 1919 suggests that the IWW was beginning to act more like a labor union and less like a revolutionary organization. Contemporary observer David Saposs contended that the IWW failed because it did not conduct systematic or methodical campaigns. But the revival of the AWIU reveals that in the postwar period the agricultural workers did plan thoughtful campaign strategies, even though members tended to revert to older and more spontaneous tactics, as the two Enid incidents indicate. Such tactics proved successful, allowing the AWIU to increase its membership and once more become the financial mainstay of the IWW.[48]

Similarly, the OWIU's decision to utilize a district approach to organization, to flood some areas with veteran union members, and to require that its officers have proven backgrounds as job delegates suggests the oil workers were moving toward more standard trade union tactics. In fact, OWIU pamphlets such as "Oil Workers!" and "Piece-Work and the Tank Builders" reveal the union's growing concern over how scientific management and piecework degraded worker control and what approaches could be used to counter it. The flyers hardly mention the One Big Union idea or the revolutionary general strike. The inclusion of tank builders in the OWIU indicates that it sought to broaden its membership base, and the obvious decision to locate the OWIU headquarters in Oklahoma City shows just how important oil workers were to the IWW's long-term strategy.

While some historians have questioned any connection between the IWW's ideas and those of the later Congress of Industiral Organizations (CIO), it seems that the OWIU was turning in the direction of CIO-style industrial unionism. But that move was thwarted, for although the new leadership was obviously more capable than previously assumed, the war had created essentially new industrial and economic conditions in Oklahoma and elsewhere and had seriously weakened the IWW financially. When new technologies appeared in the 1920s, the Wobblies were in no position to deal with them effectively.[49]

8

Decline and Fall, 1923–1930

ON February 25, 1922, a thirty-year-old Wobbly named John J. "Gasoline Slim" Leen died in an Okmulgee hospital from complications caused by his being gassed during the First World War. Leen, who had studied geology at Pennsylvania State University, was not a particularly important IWW member, but his life serves almost as a metaphor for the union. Strong in its youth, the IWW suffered serious blows from the war. Afterward, it seemed to recover, but internal weaknesses and a changing world brought it down. Tragically, many union members knew what was happening and were virtually powerless to stop it.[1]

Three major factors dealt serious blows to the IWW in Oklahoma and nationally. The first was a revived vigilantism, especially by the Ku Klux Klan, combined with renewed use of the criminal syndicalism law. The second was technological change, exemplified by the introduction of the combine harvester in the Wheat Belt and of welding in pipeline construction. The third was the serious 1924 schism within the IWW, which left the union divided and essentially directionless.

The first factor was familiar, and undoubtedly the Wobblies thought they could deal with it as they had before. As the *Industrial Pioneer* noted, the authorities failed to understand how their tactics helped the union: "[T]hey can never get it through their skulls that the spirit of solidarity thrives on persecution." But the new vigilantism was better organized, stronger, and often sanctioned by local authorities then secure in their power.[2]

The effects of technological change were more serious. The IWW was unprepared for the new technologies, which raised unemployment among the unskilled rather than the skilled. The combine harvester proved most devastating, as it effectively eliminated the union's strongest organizing base, while the increasing numbers of automobiles added to Wobbly difficulties by making workers more mobile and less accessible to the IWW.

The threat of vigilantism was not new to the Oklahoma Wobblies. The Knights of Liberty and copycat organizations were ample evidence of that, as were the litany of attacks from railroad police, newly deputized townsmen, and self-styled defenders of justice. But most of those groups were local and generally of spontaneous or situational origin. The Klan represented a more organized, brutal, and secretive opposition, which fed on the patriotic xenophobia of the postwar period. Aldrich Blake, an Oklahoma opponent of the Klan, said the Councils of Defense had encouraged spying on neighbors and vigilantism. When the war ended, the Klan often employed the same amateur agents, many of whom had belonged to the wartime American Protective League, and continued the APL's elaborate system of spies and surveillance squads. In addition, the Oklahoma KKK was one of the more powerful Klan organizations in the nation, with the fifth-largest membership overall and proportionally more members than any state except Indiana.[3]

Founded in 1919 in Oklahoma City, the Oklahoma Klan was essentially an urban organization centered around the oil industry. As a result, while the Klan generally reinforced small-town attitudes, the Oklahoma organization operated more as an extralegal arm of the business community. The relatively small immigrant and African-American populations in Oklahoma meant the Klan often focused its attacks on political radicals and union organizers, whom Klansmen called "foreign labor agitators." Significantly, the state Klan was weakest in the wheat growing areas, where a radical Populist tradition remained, and in the southeast, where tenant farmers still supported the Socialist Party.[4]

Klan growth in Oklahoma was rapid, and membership eventually reached 95,000, with the largest chapters in Tulsa and Oklahoma City. Other "Klaverns" were concentrated around established petroleum centers such as Drumright and in the emerging oil-boom towns where law enforcement was lax and vice ran rampant. Almost from its founding in Oklahoma the Klan was invited into the oil fields, as the *Oil and Gas Journal* put it, to "control brigands."[5]

While the term "brigands" usually meant bootleggers and prostitutes, it also included labor organizers, especially those of the IWW. Anti-Wobbly editorials regularly appeared in Klan newspapers and pamphlets. For example, an editorial in the *Imperial Nighthawk* claimed the IWW hated all law and order and alleged the union had an elaborate plan to conquer one state after another, with Oklahoma the next target. "Hundreds are coming into Oklahoma," the editorial declared. "You can see them loafing in the cities and tramping along the highways. Klansmen must stand shoulder to shoulder to combat the movement to trail the American flag in the dust."[6]

The exact number of direct confrontations between the IWW and the Klan is hard to determine, in part because of difficulty in precisely identifying attackers as Klansmen, especially as the IWW often indiscriminately identified all attackers as Klan members. In addition, none of the Oklahoma clashes had the scope of the Tulsa Outrage, although similar attacks did occur elsewhere. In January 1922 Klansmen tarred and feathered several OWIU organizers at El Dorado, just across the Arkansas state line, and Harold Mulkes, an IWW attorney, was kidnapped and beaten by Klansmen in Shreveport, Louisiana, when he went to defend two oil union organizers arrested there. John J. Carney faced similar threats when he arrived in Shreveport to help in the men's defense.[7]

More often the attacks involved a handful of Klansmen assaulting individual Wobblies. One such case was that of

William Bugher, an OWIU organizer, kidnapped in July 1922 near Shamrock and severely whipped and beaten by four men who more than likely were Klan members. Klansmen also clashed with AWIU organizers in Atoka and in Bald Knob, Arkansas, in 1923.[8]

The Klan also relied on more than simple thuggery. In some communities, businesses required the approval of local Klan officials before hiring men for projects. For example, the Empire Refinery of Cushing only hired men sent by candy store owner E. C. Kerns, a local "Cyclops" or Klan leader. Because many law officers were Klansmen, they singled out Wobblies for arrest and held them for several days without trial before expelling them from town. The typical charge was vagrancy, although most of the men had steady jobs and carried adequate money. The Klan membership of some district judges may also explain why the state's criminal syndicalism law, unenforced since 1920, was revived in 1923.[9]

While Oklahoma authorities never enforced the state criminal syndicalism law as thoroughly as did states such as California, the revival of the law demonstrates the growing power of the conservative incorporating forces in the state. No longer a simple response to a feared revolution, the criminal syndicalism statute—like the labor injunction—had become a legal means to harass or derail legitimate labor protest. Its greatest effect may have been in its language, which was just vague enough to have a chilling effect on many labor actions. That same vagueness also allowed for great leeway in prosecuting the law, as the IWW would discover.[10]

That the criminal syndicalism law was resurrected in 1923 is clearly no coincidence. In November 1922 Oklahoma elected as governor John C. Walton, the prolabor mayor of Oklahoma City. Walton was the candidate of the Farmer-Labor Reconstruction League. This organization was a creation of the Oklahoma State Federation of Labor, the remnants of the Socialist Party, and disaffected farmers, coal miners, and oil workers that was designed to oppose the Klan and business progressives and to capture the state Democratic Party. Once in office, Walton helped spearhead a number of major social reforms and joined four other governors in calling for the release of all federal political prisoners, including IWW members. He also incurred the wrath of the KKK when he used martial law in several counties to control Klan outrages. But when he applied martial law to the entire state, he was impeached and removed from office.[11]

Many of Walton's difficulties stemmed from the widespread control of the legislature and local governments by Klan elements. In fact, John J. Carney, the attorney who had defended both Wobblies convicted of criminal syndicalism, strongly suspected that the judge in one case was a Klansman, and Klansmen had been active in the Tri-State mining district immediately before and during the trial of the second Wobbly. Whatever the case, local officials certainly were suspicious of radicalism, and Walton's election increase their fears. As a result, in Muskogee and Miami, Oklahoma, the authorities turned to the criminal syndicalism law to stop radicalism in their communities. Their victims were two IWW job delegates: OWIU member Arthur Berg, and Homer Wear of the Metal Mine Workers Industrial Union.[12]

Berg was the first to face trial. A pipeliner and an OWIU job delegate, he was arrested by Rock Island Railroad police at Haileyville on December 27, 1922, after he had ridden a freight train from the Arkansas oil fields. Transferred to the custody of town police, he had first identified himself as "Dug Kugher from everywhere" and said he had been waiting for another train. Pittsburg County authorities then took him to McAlester, where he was convicted of vagrancy and sentenced to thirty days.[13]

After serving just over half his sentence, Berg was brought before a county judge on January 17, 1923, for arraignment on criminal syndicalism charges. The charges alleged that on December 27, 1922, Berg had sold or distributed material advocating sabotage and criminal syndicalism. The judge, a man named Treadwell, told Berg that the IWW was an illegal organization—which was untrue—and that he could face deportation because he was German. Berg replied he had been born in the United States, as were his parents. He pleaded not guilty to the charge and was bound over on a $2,500 bond—an amount that seems excessive. As he could not make the bail, Berg was returned to the county jail.[14]

IWW headquarters quickly made arrangements for Berg's defense as the trial was scheduled to start in just over a month. The IWW hired John J. Carney as Berg's attorney. A West Virginia native of Irish Catholic descent, Carney came to Oklahoma in 1894 and settled in Canadian County. Initially a stalwart in the Democratic Party, he had served as Canadian County Attorney in 1900 and 1902, and in November 1906 was elected a member of the state constitutional convention. He had been elected district judge in Canadian County in 1908 and was a Democratic candidate for Congress in 1912 before he stunned his fellow party members by announcing his conversion to socialism in late 1914. Carney became a major figure in the Oklahoma Socialist Party and regularly acted as attorney for radicals who ran afoul of the law in Oklahoma, Texas, Louisiana, and Kansas. But despite Carney's abilities, it is clear from the trial records he could do very little to prevent either Berg's or Wear's conviction, which seem foregone conclusions in retrospect.[15]

Berg's case came to trial in February 1923. Initially, Judge E. F. Lester agreed with Carney's request for a demurrer in the case on the grounds the charges were duplicitous, and he ordered County Attorney O. H. Whitt to file new charges or drop the case. The prosecutor filed a new indictment alleging that Berg knowingly and unlawfully distributed or publicly displayed material advocating criminal syndicalism or sabotage—specifically IWW membership cards, a dues book, and dues stamps—and that he was a member of an organization that advocated criminal syndicalism, namely the IWW. As the material differed little from the original information, Carney requested a second demurrer, but this time new judge, A. C. Brewster, denied the request.¹⁶

During the trial the prosecution focused almost wholly on the IWW preamble printed on Berg's membership card as proof of criminal syndicalism. The major testimony came from Dave Nowlin, the railroad police officer who had arrested Berg. Nowlin claimed to have studied IWW material, but his testimony—in which he claimed the Wobblies planned to destroy "our religion . . . our public schools, our entire form of government"—revealed he had little, if any, actual knowledge. He also said he had arrested Berg because of a state law against union organizing—a law that did not exist. Under cross-examination, he admitted Berg was unarmed when arrested and had only been walking along the tracks.¹⁷

For his part, Carney argued that the criminal syndicalism law was unconstitutional and that Berg could hardly have sold or distributed anything as he had been in Haileyville less than an hour when arrested. In addition, Carney contended that the second count of the charge—regarding Berg's IWW membership—was duplicitous and improper because belonging to the IWW was not a crime. But Judge Brewster first rejected Carney's arguments, and he then sustained an objection from the prosecution, thus preventing Carney from determining whether potential jurors were prejudiced toward the IWW.¹⁸

Carney did demonstrate that the charges against Berg stemmed from his having shown IWW pamphlets to the sons of a local restaurant owner. A policeman had taken Berg to the Star Cafe for a meal, and the young men had asked to see the material. During cross-examination, Carney also discovered that the deputy U.S. marshal who had confiscated the IWW literature had failed to obtain a search warrant. Such revelations cast serious doubt on the prosecution's case.¹⁹

Judge Brewster ignored these issues. He allowed the prosecution falsely to accuse the IWW of causing a recent train wreck and permitted testimony that was irrelevant, immaterial, and prejudicial, such as Nowlin's. All of this was intended to inflame the jury, and, under such circumstances, it seems no wonder that the jury convicted Berg on the first count, that of distributing seditious literature. They found him not guilty of belonging to an organization that advocated criminal syndicalism, which seems strange considering it was IWW literature that formed the basis of the first count.[20]

After the verdict, Brewster praised the jurors as "real Americans," then launched into a vitriolic attack on Berg. He called the Wobbly a "menace to society" who possessed "no principles" and declared that Berg had "the mark of Cain" on him. "I hope no governor would ever pardon a man out of prison for the crime you have been convicted of," the judge said.

Berg received the maximum penalty of ten years in the state penitentiary in McAlester and a $5,000 fine, plus court costs. An editorial in *Industrial Solidarity* said the penalty amounted to a life sentence, considering Berg was in his thirties and it would require thirteen years for a defendant to work out his fine at the rate of a dollar a day. "And men die early in prisons," the writer noted. Berg spent two years in prison before the Oklahoma Criminal Court of Appeals reversed his sentence. By that time he had been joined by another Wobbly, a former restaurant cook from Picher named Homer Wear.[21]

Wear was arrested while organizing a local of the IWW's Metal Mine Workers Union in the lead and zinc mines of northeastern Oklahoma. This was an especially difficult undertaking, and it remains unclear why the IWW decided to attempt a recruiting campaign there. Compared to the well-organized coal miners of southeastern Oklahoma, the lead and zinc miners of the Tri-State region remained fiercely anti-union and had earned a reputation in the far west as more-than-willing strikebreakers.[22]

For the most part native-born white Protestants, the Tri-State miners were often xenophobic and held strongly anti-Semitic and anti-Catholic attitudes. Many believed unions were un-American. The Ku Klux Klan found this fertile ground and, during the summer of 1923, regularly held parades in the mining towns and even burned a large cross at the end of Main Street in Picher. Because of their nativism, the miners remained unorganized even as lead and zinc prices fell, working conditions worsened, and absentee corporate ownership increased. Large numbers held onto long-vanished hopes of advancement, spurred in part because many miners worked ten-acre leases of their own and were effectively small-scale operators. They clung to the myth of the self-made man even as the area became "a poor man's district" with high rates of pneumonia and other respiratory diseases, four times the normal rate of typhoid, and where most children ran barefoot even in winter and few attended school.[23]

The IWW had first appeared in the region in 1906, when the Western Federation of Miners organized locals in Joplin, Missouri, and in several small mining towns in Missouri and Kansas. After leaving the IWW, the WFM found limited success, mostly organizing smelter workers in Bartlesville, Sand Springs, and Henryetta and a single miners' local in Commerce. The WFM's difficulties stemmed from the pride Tri-State miners took in being scabs and nativists, such as when they took the jobs of striking Kansas UMW members—many of them Italian and Eastern European—during the 1919 coal strike. As one miner told a social worker in 1920, "We don't want no Bolos or I.W.W.'s or labor grafters who steal the pot before the draw."[24]

Why the IWW elected to organize in this unfriendly environment remains unclear, but in the summer of 1923 Homer Wear began recruiting. Despite the hostility, the four-year veteran of the IWW found some positive responses. At a June meeting in Quapaw, a Wobbly named Oscar Citron discussed the deteriorating safety conditions and declining wages in the mines. Wear said the attendance at the meeting was good, and the men each agreed to recruit new members. Although the size of the meeting is unknown, it must have been large enough to concern local authorities, who arrested Wear on a vagrancy charge a few days later.[25]

As in Berg's case, Wear had not completed his sentence when he was charged with criminal syndicalism on August 7, 1923. The basis for the charge was a crudely written handbill that advocated violence against mine owners and ended with the statement "Boys can't be trusted; dead men don't know." The flyer was purportedly found in a trunk Wear owned, a trunk that—like Berg's possessions—had been seized without a warrant. Wear denied having written the document, and internal evidence suggests it might have been a forgery because, although riddled with grammatical errors and misspellings of simple words, more complex words were spelled correctly.[26]

Whether the handbill was a forgery did not concern Ottawa County authorities, who arraigned Wear on two counts virtually identical to those against Berg. He was then held incommunicado for thirty days and allowed neither to receive mail nor to meet with his attorney, John J. Carney. Wear did manage to send a letter detailing his plight to *Industrial Solidarity*. "I am still caged up; but am standing firm; looking forward in good faith to the future," he optimistically wrote.[27]

When the case came to trial in September, Carney again argued the law was unconstitutional, the charge duplicitous, and the evidence illegally obtained. Carney even noted that a Department of Justice agent had been allowed to see the material, again without a warrant. Interestingly enough, although the law required evidence of a defendant advocating violence, none of the prosecution witnesses testified Wear had ever done so. One witness, a water wagon driver named C. M. Smith, said he had heard Wear speak once: "He was just talking about the industrials [IWW] and the good it was to the laboring men." Wear was the only defense witness, and the local press noted that Wear's testimony as well as Carney's objections had strengthened the case for the defense. Ultimately the two-day-long trial came down to the believability of the flyer. The jury sided with the prosecution and returned a guilty verdict after only an hour of deliberation. Wear did receive a less severe sentence than Berg: six years at McAlester and a $650 fine.[28]

Both men languished in jail until 1925. In January, the Criminal Court of Appeals heard Berg's case, which had been delayed by Carney's inexplicably late filing of the appeal. While the court upheld the law's constitutionality, the judges agreed the charges were duplicitous and ruled that a charge of circulating seditious material could not be linked to another charge regarding IWW membership. The judges also concurred that Judge Brewster had improperly denied Carney the right to question potential jurors on their attitude toward the IWW. Finally, the court concluded that Dave Nowlin's testimony regarding the IWW was incompetent and prejudicial, as was the prosecution's attempt to read the IWW preamble literally. Judge Thomas A. Edwards noted the preamble could be read metaphorically, and its calls for class war could be interpreted in the same way that "Onward, Christian Soldiers" did not literally mean Christians should arm themselves for combat.[29]

The court reversed the conviction and remanded the case to the lower court, where no new charges were filed. Wear then appealed his case, which was reversed on similar grounds. *Industrial Solidarity* praised the decisions, calling the original convictions "legal lynchings" and noting that the men had been denied mail and writing privileges while in prison. After his release, Berg would go on a speaking tour to describe the brutal conditions he witnessed in McAlester, while Wear returned to northeastern Oklahoma to build up a "stake."[30]

The IWW which greeted the two men when they gained their freedom was no longer the same organization. In 1924 a series of issues had split the union into a main faction and a dissident secessionist organization, each claiming to be the true IWW. Ironically, when Wear and Berg emerged, they gravitated toward opposite camps. Berg sided with the majority faction, while Wear joined the secessionist Emergency Program. The split had a greater effect on the Oklahoma IWW than just the division of two former prisoners. It also helped wreck the oil workers' union.[31]

Although the IWW seemed on the road to recovery in the 1920s, the union still faced deep and serious internal disputes.

On the surface, the main issue revolved around the degree of centralized control in the IWW and the distribution of power within the constituent unions. The centralizing faction, associated mostly with the agricultural workers, favored the original One Big Union concept of the prewar period. In opposition were decentralizers of various stripes who worried that the General Executive Board and the AWIU had too much power and who favored giving more autonomy to constituent unions in the fashion of the AFL and later the CIO. Most opposition came from the lumberjacks and construction workers, who believed the AWIU, with its monetary clout, was too influential.[32]

But the issues went deeper. Disputes emerged over how to deal with the newly formed Communist Party, haphazard strike policies, and the advisability of class-war prisoners accepting conditional pardons. Other debate centered on a perceived toning down of rhetoric in IWW publications and on a plan to buy a headquarters building, which some more radical Wobblies saw as a betrayal of principles regarding private property. Personal animosities and frustrations complicated matters, as did the wartime approval of policies that limited the terms of officials and required them to return to the field. Ostensibly a democratic proposal to prevent oligarchic control, this rotation-in-office prevented strong action when needed, restricted the development of coherent long-term organizing tactics, and weakened the IWW at a time when it could ill afford it.[33]

The split originated in disputes over the status of the IWW political prisoners. Some prisoners favored cooperating with authorities in order to help the General Defense Committee and counter public misperceptions of the IWW. But lumberjack leader James Rowan and his supporters favored active resistance and the use of strikes on prison jobs. A number of Wobblies accepted conditional commutations in late 1923 on the condition they would be law-abiding, but eleven prisoners refused the offers of freedom. They contended that accepting the commutations amounted to admitting that their union activities were illegal. Many job delegates sided with the eleven, believing the others—mostly top officials, IWW newspaper editors, and speakers—were "prima donnas" who had betrayed the union, especially after Bill Haywood and several other prominent Wobblies defected to the Soviet Union.[34]

This dispute carried over into the question of power within the union. After financial problems emerged in 1923, the decentralizers saw the IWW as a bloated bureaucracy controlled by "piecards" and "labor fakirs" and fought back reorganizing the IWW's general administration. But the new structure—composed of a general secretary, a general organizer, and the chairs of the general organizing committees of the various industrial unions—proved unwieldy, and by 1924 the old structure was reinstated.[35]

But the debate did not die. The fall of 1924 saw the bizarre spectacle of two IWW conventions, each claiming it was the legitimate union. The mainstream faction—nicknamed the "Four Treys" because of the new headquarters at 3333 West Belmont in Chicago—constituted the largest group and included both centralizers and the majority of decentralizers, mostly AWIU members who refused to endanger the union. The dissidents, led by Rowan, called themselves the Emergency Program (EP). The EP, with headquarters in Portland, Oregon, was never very large and lasted only a few years. Still, the split was devastating because—as longtime Wobbly Fred Thompson believed—it caused large numbers of members in the middle to drop out of the union.[36]

While only four unions sided with the EP, the rift disrupted many of the smaller and financially weaker unions, such as the OWIU. The split came when the union was particularly vulnerable. Although the OWIU had made some clear gains with new groups, such as the tank builders, it remained dependent on pipeliners for the bulk of its members, whose jobs were now threatened by new technologies.[37]

Beginning in the early 1920s, several oil companies began experimenting with welded pipelines. Prior to this the threaded joints had proven the weakest parts of the line, but welded joints were as strong as the main pipes. Welding was also quicker and required far fewer men to do the work. Initially used only on smaller gasoline delivery lines, welding was first used in 1923 when Gulf Pipeline completed an eight-inch-diameter crude oil pipeline from Drumright to Yale.[38]

Some companies also turned to machinery to dig the ditches and to lay the pipes. Such machinery reduced the size of pipeline work gangs by half and did the work more quickly. Both innovations also coincided with the development of new Oklahoma oil fields that, despite widespread fears of a petroleum shortage, served to create an oil glut that caused crude prices to plummet, taking wages with them. Despite the new production, additional labor was not required, and the industry failed to absorb all those seeking work.[39]

The labor surplus resulted in a return to cutthroat competition for jobs at reduced pay. It also affected the OWIU's membership and finances. While exact OWIU membership figures do not exist, the evidence does show an increased financial strain. By October 1923 expenses exceeded revenues, and the union began that month with only $56 on hand, down from $137 in September. Relatively self-sufficient and growing throughout 1922, the OWIU twice in 1923 had to borrow $2,000 from the AWIU to stay afloat. While this might have strengthened ties between the two unions, it seems mostly to have increased animosity toward the AWIU and bolstered the decentralizers in the union.[40]

The OWIU leadership also approved actions that suggest deep divisions existed within the union. Noting that OWIU members Wencil Francik and Frank Gallagher remained behind bars, many oil workers opposed IWW members who accepted conditional commutations from President Warren Harding. Four OWIU members sent a letter protesting the practice to the OWIU headquarters in Oklahoma City. The letter was forwarded to Chicago, but the IWW leadership evidently ignored it. This suggests strong sympathy among the oil workers for the decentralizers, as do the actions of delegates to the OWIU's October 1923 convention in Oklahoma City. The delegates voted to prevent members from opening bank accounts with union funds and rejected a GEB proposal to pay officials a living wage. In contrast, the delegates also supported a controversial new Education Bureau and plans for the IWW to build its own headquarters, suggesting some sided with the centralizers and the AWIU.[41]

Other oil workers parted with the GEB over the issue of increasing initiation fees and monthly dues. The IWW constitution set the fees at extremely low levels, supposedly to make it easier for workers to join. What this actually did was deprive constituent unions of enough funds to operate efficiently. In March 1924, Ralph Colescott, a member of the OWIU's general organizing committee, wrote the GEB supporting an increase in monthly dues to one dollar. Without an increase, he argued, the constituent unions lacked funds both for effective organizing and for full participation in the decision-making process. Colescott contended OWIU proposals had been regularly ignored; and without monies to send delegates to the annual convention, which was dominated by the more prosperous AWIU, smaller unions like the OWIU had no influence over policies. "Our dues just as they are will never let us have that great privilege of representation in the general convention," he wrote.[42]

But Colescott's protests came too late. By spring the OWIU faced serious money problems, and the GEB—now back under the control of the centralizers, led by Tom Doyle—ordered the Oklahoma City OWIU headquarters closed and the union moved to Chicago, far from the oil fields. At a special OWIU meeting held on March 17, 1924, in Oklahoma City, delegates accepted the move but also urged that funds be made available to continue organizing and to extend efforts to new areas—an unrealistic hope, given the circumstances.[43]

But while financial factors certainly influenced the relocation of the OWIU headquarters, the move also suggests that the GEB made the decision in order to maintain control over the oil workers' union. For while the GEB closed down the Oklahoma City offices, it kept open much smaller OWIU branch locals in Casper, Wyoming, and in Taft and San Pedro, California. This indicates that the Oklahoma OWIU may have had more members who supported the decentralizers, even though the OWIU leadership tended to side with the GEB.[44]

Two GEB actions lend credence to this view. First, only days after the OWIU conference, the board voted to send new organizers to the Oklahoma oil fields, an action that seems redundant unless it was intended to replace existing organizers who were less sympathetic to the GEB. Second, less than a month after the move to Chicago, the OWIU announced new elections for officers, although new officers had only been chosen in October 1923. Candidates faced strict requirements: they had to have been OWIU members for at least eighteen months and to have credentials as job delegates for at least six months prior to nomination. No one could run if he had held office ninety days before the election, and candidates for the general organizing committee could not have held office for a year before the election. Current officers also had to wait a year before running again. Although such restrictions are consistent with the 1919 resolution on office holding, the speed of the new elections leads one to assume it was done to ensure officers who might be more favorable to the centralizing faction. It also gave the GEB more control over the election process itself.[45]

Whether or not that was the intent, some members may have seen it as just another betrayal by the leadership. It certainly marked the GEB's de facto abandonment of the oil workers of the Midcontinent. The feelings of betrayal were probably intensified after Charles Gibson, the OWIU delegate to the divisive 1924 IWW convention, openly sided with the AWIU and the centralizing faction against the secessionists. This action probably caused Nick Schwartz, who had been in charge of OWIU organizing in Oklahoma and Kansas, to defect to the Emergency Program faction. Probably other members simply grew disgusted with the entire squabble and dropped out of the union.[46]

The end result was the destruction of the OWIU. In September 1925, the GEB sent a new organizer to Oklahoma, and he established "branches" at post office boxes in Oklahoma City and Bristow. He experienced a notable lack of success, and in April 1926 the leadership placed the oil workers in the General Recruiting Union (GRU), a sort of catchall organization that also included Wobblies in the leather, foodstuffs, and shipbuilding industries. By July 1926 the GRU was eliminated, and, except for a few individual members, the IWW's presence in the Oklahoma oil fields ended.[47]

Compared to the OWIU, the AWIU had relative financial stability and virtual control of the IWW leadership. But that proved insufficient to ensure the agricultural workers' survival. For if technological change affected the oil workers' membership base among pipeliners, then the OWIU—given the finances—could conceivably have recruited other oil workers. This is in fact what the CIO's Oil Workers International Union later did. But the AWIU lacked such an option simply because its industry consisted solely of one kind of worker, the harvest stiff, and the union lived and died on the harvest campaign. The appearance of the combine harvester in the Wheat Belt marked the end of the harvest migratory and, with it, the end of the AWIU as an effective union.

Ironically, the increased use of the combines resulted from something the IWW had predicted: that small farms would be replaced by large-scale agribusiness operations. Oklahoma was not immune from the change. By the 1920s the size of the state's wheat farms increased an average of 81 percent, reaching a median size of 419 acres. The direct cause of the increase was a decline in wheat prices that forced farmers to produce more bushels at lower costs in order to meet their expenses.[48]

The combine—which both harvested and threshed the wheat—allowed the large-scale farms to do just that because it reduced labor costs significantly. With a combine, 5 men could do the work of 320, and those men could be locals, eliminating the need for outside help. Very quickly the combine drove the harvest stiffs from the southern plains, while the appearance of bulk storage elevators and small tractors further reduced the need for labor. As one Clinton farmer noted, combining only cost him sixteen cents per bushel, while paying for harvest hands and stationary threshing cost thirty-five cents a bushel. Many Oklahoma farmers said they expected never again to need additional help.[49]

The combine devastated the migratories. The Oklahoma State Department of Labor estimated that, between 1926 and 1927, combines had eliminated 4,000 to 4,500 harvest jobs, or almost half the total available before the war. By 1928 state wheat growers owned 4,045 combines, and the number of harvest hands supplied by the state's free employment bureaus had fallen from a wartime high of 11,296 to only 1,482. In 1932, the year it permanently closed shop, the bureau supplied only 165 men, who accepted wages of $1.25 a day. The AWIU could not organize men whose jobs no longer existed, and its membership records reflect this decline. From a high of 15,217 recruits in 1923, the numbers fell to 9,219 in 1924 and 8,507 in 1925 and worsened each year after as combine use grew in the Wheat Belt.[50]

But even before the introduction of the combine another technological innovation—the inexpensive automobile, known as the "flivver" or "tin lizzie"—also took its toll on the Wobblies. Though few in number at the start of the decade, by 1928 the automobile was the dominant mode of transportation for harvest workers in Oklahoma. The car increased the numbers of urban workers and college students looking for work, and neither group proved particularly willing to listen to IWW arguments for raising wages. These workers were often joined by impoverished small farmers from Oklahoma, Kansas, and Missouri who pooled their resources to buy a flivver and follow the harvest into the Dakotas and Canada. The result was a surplus of labor at a time when the demand declined, further shrinking wages. Making matters worse, Oklahoma and other

Wheat Belt states began enforcing laws preventing men from catching rides on freight trains. Unable to find transportation, the old hands could only watch the more mobile flivver migrants take what few harvest jobs remained.[51]

The Wobblies were powerless to change things precisely because the flivver undermined the AWIU's overall organizing strategy. Previously the Wobblies had relied on farmers having to come to the towns and recruit workers from the mass of migrants seeking jobs. With the car, prospective job hunters went directly to the farmer, eliminating whatever pressure the AWIU organizers could exert. In addition, the "honk honk hoboes" did not stay in the jungles, and the old strong-arm tactic of forcing non-Wobblies from freight cars became pointless. One Wobbly noted there was only one way to stop the flivver migrants: "A chisel through the radiator, a sledge hammer on the cylinder head, will fix them." But that comment was clearly made in frustration. There were simply too many cars.[52]

The effect on the AWIU was swift. In June 1925, three hundred AWIU members gathered in Alva to plan strategy and were able to raise wages in at least part of Oklahoma to five dollars for a ten-hour day. But it was a bitter final victory. Local authorities, such as those in Garfield County, now found it easier to deal with the Wobblies because they were among the few migrants who still rode the freight trains and congregated in towns.[53]

Wages steadily declined as the demand for harvest labor tumbled. The delegates to the 1926 AWIU spring convention in Alva voiced hopes that that year's bumper crop of 65 million bushels in Oklahoma would mean more jobs and less competition for work. They also planned to visit the auto camps and sign up the flivver migrants. "We have fought this way before, and we will fight this way again," an editorial in *Industrial Solidarity* proclaimed. But despite the bumper crops, harvest hands of all stripes encountered increasing numbers of No Help Wanted signs.[54]

In 1927, James Haney, AWIU secretary-treasurer, unrealistically expected the AWIU to have job control in the fields. By the end of the year a Wobbly writer nicknamed "Akbar" admitted that Oklahoma was lost to the union. The AWIU tried at least two other times to organize the Oklahoma harvest, but a 1929 spring harvest conference in Alva was called off because of the limited number of men in the area, and a planned 1933 meeting there apparently never took place.[55]

Robbed of its base of support in the Grain Belt, the AWIU turned its attention to other farm workers, especially immigrants in the Pacific Northwest. But without the bindlestiffs who wandered from Oklahoma to southern Manitoba, the union became a mere shadow of its former self and found success an increasingly scarce commodity.

Despite the collapse of the OWIU and the slow death of the AWIU, the IWW did pursue organizing in other parts of Oklahoma. An attempt to continue Wear's work with the Tri-State lead and zinc miners continued throughout 1926, with regular reports on conditions appearing in *Industrial Solidarity*. But the efforts were never extensive and ended after a former policeman murdered a Wobbly miner in a personal dispute. In 1927, A. S. Embree urged the GEB to recruit coal miners in Oklahoma, following some successes with UMW members in Colorado, but nothing came of his suggestion. The last IWW local organized in Oklahoma was a branch of the Construction Workers Industrial Union at Ardmore organized in 1934. It seems not to have survived the year.[56]

The 1920s proved a bitter time for organized labor in America. The incorporating forces of the open shop movement, company unionism, renewed nativism, and technological change quickly brought an end to the "new unionism" of the first two decades of the century and substantially weakened the "old unionism" of the AFL, which lost two million members between 1919 and 1933.[57]

Those forces were even more devastating for the IWW. While the union could take pride in and even draw strength from its battles with Klansmen and against the criminal syndicalism laws, these challenges distracted the IWW from the more pressing issues of technological change and the unemployment that came with it. Many of the innovations of the 1920s eliminated unskilled workers without channeling skilled workers into the ranks of the unskilled, as had happened in the late nineteenth century. This meant disaster for workers in states such as Oklahoma whose economies were based heavily in extractive industries such as oil and agribusiness. These industries were capital-intensive and not very reliant on skilled workers, so they could move rapidly toward efficiency through consolidation and cost reductions. In such cases, labor too often paid the price.

For the IWW, the price was high. The defense efforts required by the wartime prosecutions drained the union of precious funds. When the IWW began to show signs of revival, it was in the AWIU, the most vibrant branch and the one most susceptible to technological change. As a result, the AWIU's more centralized, mass-oriented tactics increasingly dominated the IWW and made the entire organization dependent on the agricultural workers' success or failure. The appearance of the combine harvester guaranteed that failure by eliminating need for the harvest stiff.

The agricultural workers' control of the IWW also contributed to the division within the union. While some historians argue that none of the issues involved were vital, that is clearly not the case. They mattered precisely because the men realized those issues were vital. The decentralizers argued that a centralized union based heavily on one industry was a vulnerable entity, while the centralizers recognized that none of the other constituent unions, with the possible exception of the oil workers', was strong enough to stand alone. When personalities and the precarious position of the organization were added, the issues came to be matters of life and death. Ironically, each group's inflexible positions worked to undermine the IWW. The decentralizers' support for limited terms of office ensured chaos and a lack of continuity in policies at all levels. The centralizers measures, such as limiting initiation fees and dues, weakened the constituent unions and left them unable to survive when the AWIU collapsed.[58]

For the OWIU, the fissures could not have come at a worse time. The restrictions on dues and fees kept the oil workers financially weak and unable to deal with the collapse in oil prices and with the replacement of pipeline gangs. Given the chance to raise its own dues and initiation fees, the oil workers' organization might have be able to focus on other aspects of the oil industry, such as refinery workers and rig builders. But the GEB tied the oil workers' hands, and that doomed the union.

As incorporating forces increased their control of Oklahoma's economy in the 1920s, the Wobblies simply failed to respond. While criminal syndicalism laws, vigilantism, and technological change took their toll, the fault was ultimately the IWW's own because the union had proven less adaptable to change than its leaders had believed. Even if it had been flexible, it was financially unable to do anything.

The bitter irony for the Wobblies was that they had seen the economic changes coming. Articles in IWW publications soberly discussed the effects of technology on unskilled workers. Wobbly writers noted how combines were eliminating harvest jobs and how new innovations such as welding and steam shovels had replaced workers in the oil and construction industries. The union offered solutions, including reduced work weeks, and urged workers to gain control of industries to implement such changes. But all of this was futile: the corporations—not the workers—owned and controlled the new technologies; and technology served shareholders and profit margins, not employees.[59]

Lacking even a modicum of workplace control and divided by competing dogmas about organization, the Wobblies rendered themselves impotent. In the end, Oklahoma was lost to the IWW as much because the Wobblies could no longer take the field as because of the strength of the incorporators. The IWW fought brilliant battles and sowed the seeds of industrial unionism in Oklahoma, but the more flexible and responsive Congress of Industrial Organizations of the 1930s reaped the benefits.

Epilogue
An Ambiguous Legacy

IN the spring of 1959 veteran Wobbly E. W. Latchem spoke on the University of Oklahoma campus in Norman. The members of an ad hoc student organization, the Norman Political Confederation, had asked the IWW's General Executive Board for a speaker, and the union had suggested Latchem, who lived in Sapulpa. Although no longer a Wobbly, he agreed to discuss the union's history and its repression during World War I.[1]

Few in the audience knew who he was, or that he had spent part of his childhood along the Little River, not far from the university. Fewer still knew that Latchem had organized railroad workers for the IWW in British Columbia, that he became an important figure in the AWO, or that he narrowly avoided conviction in Omaha on the same charges that had sent the Chicago and Wichita Wobblies to Leavenworth. Nor did they realize the old man standing in front of them had chaired the 1920 IWW convention and had written the union's formal refusal to join the Profintern, the Communist Party's international trade union organization.[2]

To them he was simply a reminder, a curiosity, or a relic of a past most Oklahomans—and Americans—had forgotten or refused to remember. The most significant indication of this collective amnesia was the absence of newspaper coverage. Only the *Tulsa Labor Press* took notice. Even the student newspaper neglected to cover his appearance. It was a far cry from the days when the mere mention of the IWW sent editors into paroxysms.[3]

Yet the fate of the old Wobbly was the fate of the IWW. Like Latchem, many members had left the organization, and as the Great Depression dawned, many sought, or drifted into, obscurity or joined mainstream labor organizations, especially the emerging CIO. Charles Krieger, who endured two and a half years in an Oklahoma jail, returned to his home in Philadelphia and became a plumber. Arthur Boose remained in the IWW but moved to the Pacific Coast, where the "Old War Horse" spent his last years in a small rented room painting portraits and hawking issues of the *Industrial Worker*.[4]

Others probably ended up like some of the Wichita Wobblies. After his release from prison, Frank Gallagher became business manager for the *Industrial Worker* and the *Industrial Pioneer*, both nearly bankrupt. After siding with the centralizers in the 1924 split, he seems simply to have vanished. Wencil Francik, who refused a commutation rather than compromise his ideals, returned to his home in Iowa, married a widow, and farmed until his death in 1947. He never again showed any interest in unionism. Oscar Gordon, the former army sergeant, also married. He worked first as a shoemaker, then, during World War II, in an aircraft factory.[5]

But others were less fortunate. The drug-addicted Albert Barr briefly returned to the IWW and to organizing oil workers. He eventually wrote a novel about his prison experiences, *Let Tomorrow Come*, but died shortly after its publication. Several others simply descended into alcoholism or remained migratories until their all-too-lonely deaths.[6]

The IWW itself barely clung to life during the 1930s, when a new era emerged for industrial unionism in Oklahoma and the rest of the United States. The union expended much of its energy trying to explain what had gone wrong. After all, it had correctly predicted the transformation of agriculture into the factory farm, and its newspapers had duly noted the effects of technology on the oil and mining industries. But the leadership was at a loss to explain why the IWW had failed to capitalize on these changes.

As early as 1925 the editors of the *Industrial Worker* argued that immigration restriction had allowed industry to absorb migratories, which was hardly the case, and that the completion of the postwar construction boom had thrown men out of work. The editorial also contended the second-hand Ford had shifted the unit of migratory labor from single men to the family, that Mexican migrants had changed the type of unskilled labor needed, and that the Communist Party had siphoned off members. But those explanations hardly accounted for the problems, especially as the Wobblies had previously thrived in such difficult circumstances.[7]

What was more important was the Wobblies' refusal to recognize that many of their problems were of their own making. While its loose structure may have aided the IWW in recruiting members, it proved disastrous in keeping members and creating stable locals. In addition, former Wobbly Joseph Murphy contended that the IWW's refusal to sign contracts meant the union could never preserve the changes it won, and its low dues and initiation fees kept locals financially strapped. The Wobblies' quasi-evangelical style of organizing meant the union too often reduced immediate strike goals to propaganda. Although effective for building worker consciousness during strikes, it was insufficient to maintain stable locals. Also, the IWW's practice of pulling out organizers and sending them elsewhere after strikes and its unwillingness to pay organizers a living wage denied locals the guidance, counsel, and sound business practices needed to survive. Joseph Ettor, who had helped lead the 1912 Lawrence strike, in 1945 speculated that the relative ease the AWIU had in gaining recruits and the fees that went along with them sidetracked the IWW from more productive areas that might have yielded more permanent results.[8]

The Wobblies also proved unwilling to work with groups that might have helped them. By rejecting cooperation with the Socialists, the Communists, and mainstream unions, the IWW maintained ideological purity, but at the expense of substantive gains. This purity also led the union to envision an idealized wageworker who embodied all working-class virtue. This unrealistic image excluded skilled workers, tenant farmers, sharecroppers, and most working women, all of whom constituted the bulk of Socialist support, especially in Oklahoma, and who could have been potential recruits for the IWW. The Wobblies' rejection of the tenant farmers proved a crucial error. Many Oklahoma tenants were impressed by Wobbly rhetoric, but in viewing the tenants only as petty landowners and little capitalists, the IWW failed to see them as they were: a degraded rural proletariat with intense grievances against the emerging business order.[9]

Conditions during the Great Depression further served to demonstrate that fact. Like the harvest workers, the tenants fell victim to technological change. The combine did not just eliminate the need for harvest workers; it also wiped out a major source of supplemental income for the tenants. The use of the Caterpillar and other types of tractors—which increased by 25 percent in Oklahoma between 1929 and 1936—helped eliminate the need for the tenants themselves.[10]

Although many responded by emigrating to the West Coast, others turned to organizing tenant unions. The Southern Tenant Farmers' Union, organized by the Socialist Party, is the best-known, but tenants in Le Flore and Sequoyah Counties, and in Fort Smith, Arkansas—the strongholds of the WCU— formed the Workingmen's Union of the World (WUW) in 1933. The WUW drew heavily on the indigenous Protestantism of the region but also relied on rhetoric derived from the Wobblies. By 1934 the WUW claimed 30,000 members in 116 locals in Arkansas and in thirty-six Oklahoma communities, although such figures are probably exaggerated. Using IWW-style direct action tactics, the WUW actively organized both the region's unemployed and the farm laborers on the potato, cotton, and corn farms that ran from the Arkansas border to Spiro and Muskogee. While it undertook no strikes, the WUW did succeed in raising pay in several localities. Even the Oklahoma tenants who migrated to California turned to unionism, with many joining the Communist-led Cannery and Agricultural Workers Industrial Union. In both cases, the IWW had lost—indeed, surrendered—important opportunities to revive its fortunes.[11]

Similarly, although the IWW had pioneered the organization of councils of the unemployed, it surrendered the high ground to others. During the 1930s, Communists helped organize eighty local Unemployed Councils with membership totaling 30,000 throughout Oklahoma. Oklahoma City alone had 7,000 members in twenty-three locals. When authorities disrupted the councils, the farmers and workers organized the Veterans of Industry of America (VIA)—led by Ira M. Finley, a former Socialist—which later absorbed the WUW. The VIA's organizers included Stanley J. Clark, the Socialist lawyer convicted with the IWW leadership in Chicago in 1918.[12]

While the IWW failed to recruit the tenants and the unemployed, it also found that its sacred banner of industrial unionism had been taken up in Oklahoma—as elsewhere—by the Congress of Industrial Organizations. When the CIO appeared, it inherited a divided working class, a labor movement still devoted to a declining craft unionism, and a society lacking many common working class institutions. But it also inherited the egalitarian traditions and syndicalist ideals of the IWW, as well as the new unionism movement of several unions within the turn-of-the-century AFL. Yet the CIO succeeded where the IWW had floundered, in part because the newer organization did not repeat the Wobblies' mistakes, such as refusing to sign time contracts and rejecting political action. The CIO's decentralized structure also meant the entire organization was not dependent on the success or failure of any one branch, as the IWW had been on the AWIU.[13]

More important, the CIO grew from the strong and stable industrial unions of the AFL. In Oklahoma, the CIO found success in the traditionally militant and pro-union coal mines and tenant farms of the southeast and in the packinghouses of Oklahoma City. It also gained ground in the less enthusiastic oil fields and among the formerly anti-union and nativist lead and zinc miners of the tristate. When officially chartered in June 1937, the Oklahoma CIO counted among its members 7,500 colliers; 8,000 oil field and refinery workers; 3,000 metal miners and smelter workers; 2,000 glass workers; and 200 journeyman tailors. Combining the large membership base with long-standing stable business practices, the CIO possessed two factors necessary for survival, factors the IWW lacked.[14]

The unions that composed the Oklahoma CIO also drew on the state's radical political traditions and made use of the electoral process to gain support. Its leaders included J. Luther Langston, Edgar Fenton, and Victor Purdy, all either ex-Socialists or former leaders of the Oklahoma Federation of Labor, and L. N. Sheldon, who had canvassed the state for the Non-Partisan League. Constituent unions, such as the oil workers', used the courts and the injunction process against employers, and the semi-official CIO paper, *Oklahoma Labor*, regularly editorialized for political reform favorable to labor, tactics that were anathema to the Wobblies. But the paper also carried attacks against company unions, against the craft policies of the AFL, and in favor of industrial unionism in terms the IWW would have recognized. Victor Purdy, in one article, also praised the IWW as a forbearer of the CIO.[15]

Although no CIO unions have direct ties to the IWW, the oil workers' union comes the closest, especially in Oklahoma. Certainly some members of the revived International Association of Oil Field, Gas Well, and Refinery Workers had been Wobblies, especially those in the Drumright area, where pipeliners were often "two-card men," with membership in both the AFL and the IWW. Others clearly were influenced by the IWW. Among the two-card men was probably Henry W. Wier, later an official with the International Association, who worked as a pipeliner near Cushing in 1917. Another oil worker and CIO organizer, known as "Walter Strong," had taken part in the Green Corn Rebellion, while Covington Hall said many veterans of the Brotherhood of Timber Workers who helped organized the IWW oil workers' union were also active in the CIO. Wobbly influence can also be seen in the name the union adopted, the Oil Workers International Union, or OWIU-CIO, and in the song chosen as the union's official anthem—the old IWW song "Solidarity Forever," written by Ralph Chaplin.[16]

Even before the birth of the CIO, the oil workers' union made gains in the Midcontinent fields. Its first major Oklahoma local was organized in Seminole in 1929. Other locals were organized in areas familiar to the Wobblies: Cushing, Drumright, Healdton, and Tulsa. Until the Franklin Delano Roosevelt administration, however, the union remained weak, numbering only 300 to 400 members as late as 1932. But Section 7a of the National Industrial Recovery Act proved a boon for the oil workers and helped them obtain something the IWW never had: signed contracts with the oil companies. By using selected strikes and appealing to the Petroleum Labor Policy Board, the oil workers in 1933 reached agreement in twenty strikes, gaining eight signed contracts, and achieved contracts or mediation in fifteen threatened strikes. One major victory was a national contract with Sinclair Oil; and by 1936 the union had gained contracts with other Oklahoma oil producers, including Cities Services Oil and Gas Corporation, Indian Territory Illuminating Oil Company, Champlin Petroleum Company, and Deep Rock Petroleum. As a result of the successes, union membership grew to 75,000 members by 1936, with District Five, which included all of Oklahoma, the largest district.[17]

This probably seemed a bitter blow to the IWW. Wobblies had opposed the NIRA and the later Wagner Act on the grounds that they constituted undue interference in labor-employer relations, a position they ironically shared with the National Association of Manufacturers, the Allied Industries of Oklahoma, and the state's various chambers of commerce. It also demonstrated just how far the IWW had drifted from the mainstream of American labor and from the rank and file, and how little it understood the necessity for political as well as economic action.[18]

Not everything was successful for the OWIU-CIO, however. As late as 1941, the union had yet to organize to the major corporations, such as Standard Oil, and a lengthy and indecisive strike at Tulsa's Mid-Continent Refinery demonstrated the resolve of oil companies to wreck the union. Membership dropped to 20,000, and the strike left the union's defense fund and resources exhausted until the United Mine Workers provided a $50,000 loan. The union revitalized in 1940 only after insurgents from District Five successfully pushed through a new program based on a concept familiar to the Wobblies— worker control—and after World War II and a nationwide postwar strike in 1945 brought workers at Standard, Pure, Shell, and Texaco under the CIO banner.[19]

Like the IWW, the OWIU-CIO faced attacks from Oklahoma authorities and newspapers, which called it "communistic," just as they had the Wobblies. In addition, state authorities used the National Guard to end the Mid-Continent Refinery strike, while legislators considered reviving and amending the old anti-IWW criminal syndicalism law to counter the sit-down strike. Pioneered by the Wobblies, the sit-down was first used in Oklahoma by the oil workers during a 1932 strike at Drumright's Tidewater refinery. In 1939 Rogers County Rep. Tom Kight introduced an amendment outlawing the sit-down strike and gained the support of the Allied Industries of Oklahoma, the Anti-Communist League of America, and the Oklahoma City and Tulsa press. But lobbying by labor officials, including C. H. Chaffin of the OWIU-CIO, led to the passage by the state house and senate of differing watered-down measures. House leaders then accepted the senate proposals, which failed to garner enough votes for approval. The Wobblies' rejection of politics would never have allowed such lobbying, and it seems reasonable to assume a different fate for the law without the CIO's action.[20]

The criminal syndicalism law itself died shortly thereafter. But it was not the Wobblies who killed it—it was their archrivals, the Communists. In August 1940, Oklahoma City police raided the Progressive Bookstore and arrested several actual and suspected CP members, including the party's state secretary, Robert Wood, and his wife, Ina. They and two others— city CP secretary Alan Shaw and Eli Jaffe, a Brooklyn graduate student—were convicted, but the Oklahoma Criminal Court of Appeals twice reversed the sentences, arguing that, although the law was constitutional, it was applied unconstitutionally. The ruling effectively made the law unenforceable.[21]

In addition, unlike the cases involving Wobblies, many state and national newspapers, such as New York's *PM*, the *Saint Louis Post-Dispatch*, and Wisconsin's *Madison Capital-Times*, attacked the law as a violation of the First Amendment and American principles of free speech. Editorial writers nationwide compared the law unfavorably to the tactics of the Nazis, although E. K. Gaylord's *Oklahoma City Times* and the *Daily Oklahoman*—whose editorial writers included the bill's original author, Luther Harrison—defended it.[22]

But the political atmosphere of the late 1930s and early 1940s differed from that of World War I and the 1920s. The time was ripe for industrial unionism. The Great Depression had weakened, although not fatally, the power of the incorporating forces, and the election of Franklin Roosevelt both changed organizing conditions and produced a federal government less willing to give military and judicial support to employers. Even Congress, through the NIRA's Section 7a and the Wagner Act, had told business leaders that they must recognize unions. State governments too rejected the use of the National Guard as strikebreakers, and some began to dismantle their World War I-era antilabor laws, although Oklahoma was not immediately among them.[23]

Yet while the times had changed, the IWW had not, and therein lies the major explanation for its decline. Born in the labor struggles of the western industrial frontier, the Wobblies utilized strategies and tactics appropriate for the migratory workers of that region. The short strike, sabotage, rejection of political action, and refusal to sign contracts were useful where political, judicial, and social institutions were either local or nonexistent, as they were in most of Oklahoma before the First World War and shortly after. These tactics helped build class-consciousness and instilled a fighting spirit in the workers as they dealt with the emerging structure of an industrial society that refused to recognize their rights.[24]

The IWW utilized these strategies and tactics long after their effectiveness had declined. Without methods that could deliver concrete results, the IWW was unable to recruit nonmigratory workers in mass production industries. Its organizational approach limited the IWW only to migratories such as the harvest hands and the pipeline workers of Oklahoma. As a result, the union's decline was not so much that the "Bummery" captured the IWW, as many historians have argued, but that the IWW captured the Bummery.[25]

These unskilled workers—far from being the franc tireurs of the revolution—proved unable to sustain a union either financially or practically. Their day-to-day struggle meant they lacked the monies to fund a healthy union, and they lacked the time, education, physical energy, and even emotional security to devote to a union. Moreover, these men were actually transitional figures in the industrial process. Created by the partial rationalization of wheat agriculture and oil production, they were the most easily replaced when improved technology appeared. There was little the Wobblies could do in respect to the wheat harvest, but the IWW's inability to change its tactics cost it an opportunity in the petroleum industry. The more flexible CIO took advantage of the new economic reality by appealing to the settled, nonmigratory worker and by exploiting unionism's limited position as a countervailing force in both the economic and political spheres.[26]

Still, the importance of the IWW in Oklahoma should not be discounted. The Wobblies articulated grievances against the new economic order, which showed otherwise powerless groups such as agricultural laborers and tenant farmers that they did not have to rely solely on the political process to produce change, that they themselves had the power to improve their own lives. By demonstrating that industries such as oil could be organized industrially, the IWW prepared the way for the CIO. Finally, in challenging legal structures such as the federal conspiracy statutes and the Oklahoma criminal syndicalism law, the Wobblies helped establish precedents important for the defense of both labor organizing and the rights of free speech and a free press.

The IWW's tragedy, then, is that it failed itself to benefit from the struggles of its members. But as Dave Archibald, a veteran Oklahoma UMW member, would later say of the Wobblies: "Don't let anybody ever tell ya' they wasn't all right too. They's good people; they done a lot of good deeds in that Western country."[27]

Notes

Introduction

1. A discussion of the hostile charivari is in Bertram Wyatt-Brown, *Honor and Violence in the Old South* (New York: Oxford Univ. Press), 1986, 188-204; National Civil Liberties Bureau, "The 'Knights of Liberty' Mob and the IWW Prisoners at Tulsa, Oklahoma, November 9, 1917" (New York: National Civil Liberties Bureau, February 1918), 6-7 (hereafter NCLB, "'Knights of Liberty' Mob"); *Tulsa Daily World*, November 10, 1917; *Tulsa Democrat*, November 11, 1917.

2. The origin of the term "Wobbly" for IWW members is unclear. Several stories, mostly apocryphal, exist. For a full discussion, see Archie Green, *Wobblies, Pile Butts and Other Heroes: Laborlore Explorations* (Urbana: Univ. of Illinois Press, 1993), esp. Chapter 3, "The Name *Wobbly* Holds Steady," 97-138.

3. The history of radicalism, especially of the Socialist Party, in Oklahoma is well treated in James Green, *Grass-Roots Socialism: Radical Movements in the Southwest, 1895-1943* (Baton Rouge: Louisiana State Univ. Press, 1978); Garin Burbank, *When Farmers Voted Red: The Gospel of Socialism in the Oklahoma Countryside, 1910-1924* (Westport, Conn.: Greenwood Press, 1976); Worth Robert Miller, *Oklahoma Populism: A History of the People's Party in the Oklahoma Territory* (Norman: Univ. of Oklahoma Press, 1987); and John Thompson, *Closing the Frontier: Radical Response in Oklahoma, 1889-1923* (Norman: Univ. of Oklahoma Press, 1986). With the exception of Green, however, all these works are weak in their discussion of the role of industrial labor, and even Green downplays the role of the IWW. See J. Green, *Grass-Roots Socialism*, 193-204, 369-70. On the subject of the incorporation of the United States and of the West and, in particular, on the violence it produced see Alan Trachtenberg, *The Incorporation of America: Culture and Society in the Gilded Age* (New York: Hill and Wang, 1982), 3-10; and Richard Maxwell Brown, "Law and Order on the

American Frontier: The Western Civil War of Incorporation," in *Law for the Elephant, Law for the Beaver: Essays in the Legal History of the North American West*, ed. John McLaren, Hamar Foster, and Chet Orloff (Regina, Saskatchewan: Canadian Plains Research Center University of Regina, 1992), 74-89. The amount of space two modern chroniclers of the IWW devote to the AWO demonstrates the lack of attention paid to that IWW branch. Melvyn Dubofsky devotes just nine pages to the union, while Philip S. Foner is only slightly more generous, giving the AWO a thirteen-page chapter and a handful of other references. Both discuss the AWO in fairly general terms and emphasize its leadership and its disputes with the rest of the IWW. See Melvyn Dubofsky, *We Shall Be All: A History of the Industrial Workers of the World*, 2d ed. (Urbana: Univ. of Illinois Press, 1988), 313-18, 343-45; Philip S. Foner, *History of the Labor Movement in the United States*, vol. 4, *The Industrial Workers of the World, 1905-1917* (New York: International Publishers, 1965), 473-85 (hereafter Foner, *IWW History*). The only study devoted exclusively to the AWO is Stanley P. Fast, "The Agricultural Workers Organization and the Harvest Stiff in the Midwestern Wheat Belt, 1915-1920" (master's thesis, Mankato State University, 1974). Fast, however, emphasizes union activity in the northern Great Plains and spends little time on either Oklahoma or Kansas.

4. The effects of this period on workers are well discussed in David Montgomery, *Workers' Control in America: Studies in the History of Work, Technology, and Labor Struggles* (New York: Cambridge Univ. Press, 1979); Trachtenberg, *Incorporation of America*, 21-22.

5. William Preston, "Shall This Be All? U.S. Historians versus William D. Haywood et al.," *Labor History* 12 (summer 1971), 438-39; David Brody, "Labor Movement," in *Encyclopedia of American Political History: Studies of Principal Movements and Ideas*, ed. Jack Greene (New York: Scribners, 1984), 712.

6. The development of workers' republicanism in the pre-Civil War era is best discussed by historian Sean Wilentz, while historical sociologist Kim Voss has noted the republicanism inherent in the ideology of the Knights of Labor. Evidence for this as it pertains to Oklahoma can be found in letters and articles, mainly by the Knights' lobbyist Ralph Beaumont, in the union's own *Journal of United Labor* and the *Oklahoma War Chief*, based in Caldwell, Kansas, and edited by Samuel Crocker, a Boomer and a Knight. Sean Wilentz, *Chants Democratic: New York City and the Rise of the American Working Class, 1788-1850* (New York: Oxford Univ. Press, 1984), esp. 14-15, 92-97, 113; Kim Voss, *The Making of American Exceptionalism: The Knights of Labor and Class Formation in the Nineteenth Century*

(Ithaca: Cornell Univ. Press, 1993), 26-27, 80; W. L. Couch, John Blackburn, E. S. Wilcox, and Samuel Crocker to Terence V. Powderly, December 4, 1885, in *Journal of United Labor*, January 10, 1886; Terence V. Powderly, "Who Owns the Coal Fields?" ibid., March 10, 1888, April 4, 1889; *Oklahoma War Chief*, July 1, 8, 1886.

7. The concept of "business progressives" is borrowed from George Brown Tindall and was applied to Oklahoma by James Scales and Danney Goble. While Tindall sees them as essentially southern and Democratic, business progressives—usually Republicans—existed also in the West. In fact, as Scales and Goble have pointed out, those southern Democratic business progressives who came to Oklahoma functioned much like their Republican incorporator counterparts in the West. George Brown Tindall, *The Emergence of the New South, 1913-1945* (Baton Rouge: Louisiana State Univ. Press, 1967), 219, 224; James R. Scales and Danney Goble, *Oklahoma Politics: A History* (Norman, Univ. of Oklahoma Press, 1982), 52, 64-65; J. Green, *Grass-Roots Socialism*, 369-70; Brown, "Law and Order on the American Frontier," 74-79; J. Thompson, *Closing the Frontier*, 80.

8. The concept of Oklahoma Territory as a "businessman's frontier" comes from Norman L. Crockett, although this idea also readily applies to Indian Territory as well. Sociologist Ellen I. Rosen's "Socialism in Oklahoma" is also a good overview of how corporations came to control Oklahoma. Norman L. Crockett, "The Opening of Oklahoma: A Businessman's Frontier," *Chronicles of Oklahoma* 55 (spring 1978): 85-95; Ellen I. Rosen "Socialism in Oklahoma: A Theoretical Overview," *Politics and Society* 8 (1978): 119-23; Arrell Morgan Gibson, *Oklahoma: A History of Five Centuries*, 2d ed. (Norman: Univ. of Oklahoma Press, 1981), 158, 163-64, 169, 171; Jerome O. Steffen, "Stages of Development in Oklahoma History," in Anne Hodges Morgan and H. Wayne Morgan, eds., *Oklahoma: New Views of the Forty-Sixth State* (Norman: Univ. of Oklahoma Press, 1982): 25-26.

9. Historian Richard White goes even further, arguing the railroads provided access to eastern and European markets and created and developed the market for Great Plains wheat. The concept of a wage-worker's frontier created by the railroads is Carlos Schwantes's, but the late Arrell Morgan Gibson noted this mobility was common to most Oklahoma economic enterprises, including farming and oil drilling, and prevented the development of a "sense of place" in the region. Curiously, while the literature on the coal mines of Indian Territory is extensive, none of it has examined the industry as an example of an urban-industrial

frontier created by the railroads. Crockett, "Opening of America," 88-90; Gibson, *Oklahoma: A History*, 158-60, 173; Arrell Morgan Gibson, "Oklahoma, Land of the Drifter: Deterrents to Sense of Place," *Chronicles of Oklahoma* 64 (summer 1986): 5-13; Frederick Lynne Ryan, *The Rehabilitation of Oklahoma Coal Mining Communities* (Norman: Univ. of Oklahoma Press, 1935), 26-7, 32-5; Carlos A. Schwantes, "The Concept of the Wageworkers' Frontier: A Framework for Future Research," *Western Historical Quarterly* 18 (January 1987): 41, 43, 48-49; Richard White, *"It's Your Misfortune and None of My Own": A New History of the American West* (Norman: Univ. of Oklahoma Press, 1991), 243-45, 257; John Womack, Sr., *Norman: An Early History, 1820-1900* (Norman: Womack, 1976), 24-26; *Journal of United Labor*, April 4, 1889; August 2, 1889; September 19, 1889; *Industrial Union Bulletin*, February 27, 1909.

10. Dubofsky, *We Shall Be All*, 81, 149-50; Ryan, *Rehabilitation*, 65, 69; Schwantes, "Wageworker's Frontier"; J. Thompson, *Closing the Frontier*, 23-24; Steffen, "Stages of Development," 27; Steve Sewell, "Amongst the Damp: The Dangerous Profession of Coal Mining in Oklahoma, 1870-1935," *Chronicles of Oklahoma* 70 (spring 1992): 67-70; Gene Aldrich, "A History of the Coal Industry in Oklahoma to 1907" (Ph.D. dissertation, University of Oklahoma, 1952), 72-73.

11. Clark Kerr and Abraham Siegel, "The Interindustry Propensity to Strike: An International Comparison," in *Industrial Conflict*, ed. Arthur Kornhauser, Robert Dubin, and Arthur M. Ross (New York: McGraw-Hill, 1954), 191-96; Dubofsky, *We Shall Be All*, vii, 150. See also Dubofsky, "The Origins of Western Working Class Radicalism," in *The Labor History Reader*, ed. Daniel J. Leab (Urbana: Univ. of Illinois Press, 1985), 230-53; J. Thompson, *Closing the Frontier*, 82, 88.

12. Evidence collected by Jonathan Garlock suggests, however, that two local assemblies were in the coal fields as early as 1880. See Jonathan Garlock, *Guide to the Local Assemblies of the Knights of Labor* (Westport, Conn.: Greenwood Press), 1982, 402; Federal Writers Project of Oklahoma, Works Progress Administration, *Labor History of Oklahoma* (Oklahoma City: A. M. Van Horn, 1939), 4-5, 6-12.; Ryan, *Rehabilitation*, 27-28, 32; Aldrich, "History of the Coal Industry," 75-85. For a fuller discussion of the twin territories' Knights of Labor, see Nigel Anthony Sellars, "Miners, Farmers, and Politicians: The Knights of Labor in Oklahoma, 1882-1894" (paper presented at the Oklahoma Regional Conference of Phi Alpha Theta, University of Oklahoma, Norman, March 27, 1993); Miller, *Oklahoma Populism*, 95; *Daily Oklahoma State Capital*, June 7, July 9, 1894; *Daily Oklahoma Times-Journal*, July 9, 1894; *Guthrie Daily Leader*, July 11, 1894.

13. As only intermarried whites had citizenship rights, all other whites were left powerless, which rendered electoral politics of any sort irrelevant before statehood. But the miners were luckier than the tenant farmers, who lacked the right to own land or even to send their children to school. However, all whites were effectively "guest workers" in the Indian nations and had to pay a license fee to remain. Because the coal companies paid the fees directly to the tribe, the corporations wielded a powerful weapon over the workers. The companies withdrew the fees during the 1894 strike, which became the excuse for using federal troops to expel the miners. See J. Thompson, *Closing the Frontier* 82, 88.

14. These factors for defining syndicalism are borrowed from English historian Eric Hobsbawm. While historian David Shannon has noted American syndicalism was a hodgepodge of ideas only roughly parallel to the European version, his description of its attributes virtually mirrors Hobsbawm's. One of the first writers to call the IWW syndicalist was journalist John Graham Brooks. But IWW editors, such as Ben H. Williams, generally rejected the label as too European and preferred the term "industrial unionism." The IWW's calls for "One Big Union," for a general strike, and for workers' control of industry are clearly syndicalist. Eric Hobsbawm, *Workers: Worlds of Labor* (New York: Pantheon, 1984), 273-74, 276-77; David A. Shannon, *The Socialist Party of America: A History* (New York: Macmillan, 1955), 38; John Graham Brooks, *American Syndicalism: The IWW* (New York: Macmillan, 1913); *Solidarity*, September 14, 1912.

15. The Choctaw tribal government and laws, however, made it difficult to organize cooperative stores. "Romulus," Lehigh, Indian Territory, to the editor, May 28, 1888, *Journal of United Labor*, June 16, 1888; Fred Reitner, Lehigh, Indian Territory, to editor, undated letter, ibid., September 20, 1888; Walter Kerr to editor, undated letter from Lehigh, Indian Territory, ibid., January 10, 1889. On the Knights as an alternative workers' culture, see Mike Davis, *Prisoners of the American Dream: Politics and Economy in the History of the U.S. Working Class* (London: Verso, 1986), 30-31.

16. Aldrich, "History of the Coal Industry," 71-72; Federal Writers Project of Oklahoma, *Labor History*, 5-6; John Barnhill, "Triumph of Will: The Coal Strike of 1899-1903," *Chronicles of Oklahoma* 61 (Spring 1983): 83-85. On the UMW in Oklahoma politics, especially in the Socialist Party, see J. Green, *Grass-Roots Socialism*, 193-204. For the political activities of the Oklahoma labor movement as a whole, see Keith L. Bryant, "Labor in Politics: The Oklahoma State Federation of Labor during the Age of Reform," *Labor History* 11 (summer 1970): 259-76.

17. *Proceedings of the Sixth Annual Convention of the Oklahoma State Federation of Labor, 1909*, Tulsa (Oklahoma City: Warden-Ebright Printing, 1909), 23, 41; *Official Yearbook and Proceedings of the Seventh Annual Convention of the Oklahoma State Federation of Labor, 1910*, Chickasha, (Oklahoma City: Allied Printing Trades Council, 1910), 91; *Official Yearbook and Proceedings of the Eighth Annual Convention of the Oklahoma State Federation of Labor, 1911*, Bartlesville (Oklahoma City: Allied Printing Trades Council, 1911), 93, 105. Copies in the Oklahoma State Federation of Labor Collection, Western History Collections, University of Oklahoma, Norman (hereafter OSFL Collection, WHC/OU), Boxes 70 and 88.

18. Preston, "Shall This Be All?" 440; David Saposs, *Left-Wing Unionism: A Study of Radical Policies and Tactics* (New York: International Publishers, 1926), 133-37.

Chapter 1. Impossibilists, Reformers, and Labor Fakirs

1. Foner, *IWW History*, 29-31; Dubofsky, *We Shall Be All*, 81-87; the Industrial Union Manifesto of January 1905 is reprinted in the *Miners' Magazine*, January 26, 1905, and *Voice of Labor*, March 1905.

2. Actually, the AFL did attempt to organize unskilled laborers and other workers who lacked national unions through its Federal Labor Unions program. But the FLUs never organized extensively and never represented more than 7 percent of AFL members, and only 1.3 percent by 1910. By 1911, Oklahoma had five such unions, at Wilburton, Bartlesville, Henryetta, Durant, and Oklahoma City; Philip S. Foner, *History of the Labor Movement in the United States, vol. 3, The Policies and Practices of the American Federation of Labor, 1900-1909* (New York: International Publishers, 1964), 198-99, 434 (hereafter, Foner, *Labor History*); William M. Dick, *Labor and Socialism in America: The Gompers Era* (Port Washington, N.Y.: Kennikat Press, 1972), 89-92; *Proceedings of the Eighth Convention of the Oklahoma State Federation of Labor, 1911*, OSFL Collection, WCH/OU, Boxes 70 and 88, 93; Dubofsky, *We Shall Be All*, 19-38, 57-60, 66-76; Foner, *IWW History*, 401, 413-38; "Read and Consider," *Voice of Labor*, February 1905; *American Labor Union Journal*, April 16, April 28, May 5, 1904. WFM locals in the tristate can traced in several issues of the *Miners' Magazine*, including January 14, May 26, 1904, and February 9, 1905.

3. Foner, *IWW History*, 29-31; Dubofsky, *We Shall Be All*, 81-87; *Miners' Magazine*, January 26, 1905; *Voice of Labor*, March 1905.

4. Hagerty did tell the convention delegates a story about a railroad worker in Indian Territory who failed to understand the class struggle. The man sat under a water tank to avoid the summer heat and scratched

himself vigorously. After a while, he would remove a biting insect, examine the creature carefully, then put it back in the same place. When another worker asked why the man did not kill the insect, the first man replied, "Why, that wasn't the one that was biting me." Hagerty said too many workers were like that man: putting a parasite back because it was not biting them at just that moment. The story appeared in *Miners' Magazine*, August 10, 1905; Robert E. Doherty, "Thomas J. Hagerty, the Church, and Socialism," *Labor History* 3 (winter 1962): 40, 43; J. Green, *Grass-Roots Socialism*, 30; *Proceedings of the National Convention of the Socialist Party, Chicago, Ill., May 1 to May 6, 1904* (Chicago: Socialist Party of America, 1904), 45; *Miners' Magazine*, March 23, 1905, 5; Paul F. Brissenden, *The IWW: A Study of American Syndicalism* (New York: Russell and Russell, 1957), 68.

5. *Industrial Worker* (Joliet, Ill.), February, March, April 1906; Arrell Morgan Gibson, *Wilderness Bonanza: The Tri-State District of Missouri, Kansas, and Oklahoma* (Norman: Univ. of Oklahoma Press, 1972), 227, 230.

6. Montgomery, *Workers' Control*, 91–92; Steffen, "Stages of Development," 19, 24; *Miners' Magazine*, April 27, 1905, 3; *Unionist*, April 27, May 1, June 23, 1905; Sister M. Magdalen Reinhart, O.S.B., "The Open Shop Movement in Oklahoma" (master's thesis, University of Oklahoma, 1938), 1, 4.

7. Clashes between progressive reformers and labor, especially the IWW, were common in the West, as the business and professional elements were attempting both to seize power from rural and working-class groups and to impose their standards of morality and discipline on working-class leisure activities, such as drinking. Herbert G. Gutman, "The Workers' Search for Power: Labor in the Gilded Age," in Herbert G. Gutman, *Power and Culture: Essays on the American Working Class*, ed. Ira Berlin (New York: Pantheon Books, 1987), 72–74, 91–92; Scales and Goble, *Oklahoma Politics*, 64–5; William D. Rowley, "The West as Laboratory and Mirror of Reform," in *The Twentieth-Century West: Historical Interpretations*, ed. Gerald D. Nash and Richard W. Etulain (Albuquerque: Univ. of New Mexico Press, 1989), 345.

8. The term "new unionism" derives from the British labor movement of the late 1800s but seems to have first been applied to the United States by Andre Tridon. Articles on the union label and legislative efforts were regular features of Oklahoma AFL papers such as the Oklahoma City *Oklahoma Labor Unit*. One important legislative achievement was the banning of "job sharks," who sold jobs for a fee, which they often pocketed without delivering the job. This was a problem the IWW fought

in the lumber camps of the Northwest. Andre Tridon, *The New Unionism* (New York: B. W. Heubsh, 1913); Milton Ernest Asfahl, "Oklahoma and Organized Labor" (master's thesis, University of Oklahoma, 1930), 126; Montgomery, *Workers' Control*, 48, 64, 72, 93, 100; Scales and Goble, *Oklahoma Politics*, 63-64; James R. Green, *The World of the Worker: Labor in Twentieth Century America* (New York: Hill and Wang, 1980), 35-38; *Labor Amalgamator*, January 1904; *Industrial Democrat*, April 9, 1910.

9. "S.," Shawnee, Oklahoma Territory, to editor, *Industrial Worker* (Joliet, Ill.), March 1906.

10. *Industrial Worker* (Joliet, Ill.), February, September 1906; Garin Burbank, "Socialism in an Oklahoma Boom-Town: 'Milwaukeeizing' Oklahoma City," in *Socialism in the Cities*, ed. Bruce M. Stave (Port Washington, N.Y.: Kennikat Press, 1975), 102-103; Lee Peeler, ed., *The Standard Blue Book of Oklahoma, 1910-1911* (Oklahoma City: A. J. Peeler and Co., 1909), 100, 106, 108, 165; *Labor Amalgamator*, December 1903.

11. On racial attitudes of native-born workers in the tristate, see Charles Morris Mills, "Joplin Zinc: Industrial Conditions in the World's Greatest Zinc Center," *Survey* 45 (February 5, 1921), 657-58. For evidence that some Wobblies were themselves racists, Len De Caux describes an encounter with a Southern hobo named "Red" who supported both the IWW and the Klan. See De Caux, *Labor Radical: From the Wobblies to the CIO, a Personal History* (Boston: Beacon Press, 1970), 82-83; Grover H. Perry, "The Revolutionary IWW" (Cleveland: Industrial Workers of the World Publication Bureau, July 1916), copy in Redmond S. Cole Collection, Western History Collections, University of Oklahoma, Norman (hereafter Cole Collection, WHC/OU), 11; Donald Ralph Graham, "Red, White, and Black: An Interpretation of Ethnic and Racial Attitudes of Agrarian Radicals in Texas and Oklahoma, 1889-1920" (master's thesis, University of Saskatchewan-Regina, 1973), iv, 248; Ann Schofield, "Rebel Girls and Union Maids: The Woman Question in the Journals of the AFL and IWW, 1905-1920," *Feminist Studies* 9 (summer 1983): 335-38, 341-46, 348, 349, 351, 354; Elizabeth Jameson, "Imperfect Unions: Class and Gender in Cripple Creek, 1894-1904," in *Class, Sex, and the Woman Worker*, ed. Milton Cantor and Bruce Laurie (Westport, Conn.: Greenwood Press, 1977), 171-72, 174-76, 185; *Industrial Worker* (Joilet, Ill.), April 6, 1907.

12. *Warden-Ebright's Oklahoma City Directory, 1906-1907* (Oklahoma City: Warden-Ebright Printing, 1906), copy in Library, Oklahoma Historical Society, Oklahoma City; "The Resolution of Oklahoma City," *Industrial Worker* (Joliet, Ill.), August 1906; ibid., July 1906; Ethel E. Carpenter, "The Farmer and His Hired Hand," ibid., April 1906.

13. Meredith Tax, *The Rising of the Women: Feminist Solidarity and Class Conflict, 1880–1917* (New York: Monthly Review Press, 1980), 129–31, 143; Schofield, "Rebel Girls"; William D. Haywood to Mrs. Elmer J. Buse, in Melvyn Dubofsky, ed., *United States Department of Justice Investigative Files, Part I: The Industrial Workers of the World* (Bethesda, Md.: University Publications of America, 1989), Reel 4, f. 0989 (hereafter *DOJ/IWW Microfilm*). This microfilm set includes material from National Archives Record Group 60, Casefiles 186701 and 189152. Street's reply, on f. 0990, is well-known and was published in Daniel T. Hobby, "'We Have Got Results': A Document on the Organization of Domestics in the Progressive Era," *Labor History* 17 (winter 1976): 104. The Post Office intercepted the original letter in 1917 and forwarded it to the Department of Justice.

14. Dick, *Labor and Socialism*, 83–84, 89–92; J. Green, *Grass-Roots Socialism*, 36–37, 225–26; James Weinstein, *The Decline of Socialism in America, 1912–1925* (New Brunswick, N.J.: Rutgers Univ. Press, 1984), 37; Ira Kipnis, *The American Socialist Movement, 1897–1912* (New York: Columbia Univ. Press, 1952), 392.

15. Joseph Conlin, *Bread and Roses, Too: Studies of the Wobblies* (Westport, Conn.: Greenwood Press, 1969), 83–89.

16. Dick, *Labor and Socialism*, 34–36, 184; Dubofsky, *We Shall Be All*, 12–13; Grover H. Perry, "How Scabs Are Bred," in Perry, "*Revolutionary IWW*," 25; "Union Scabs—and Others," is reprinted in Oscar Ameringer, *If You Don't Weaken: The Autobiography of Oscar Ameringer* (New York: Henry Holt and Co., 1940), 189–92.

17. At statehood, the UMW had 7,000 state members, and all other unions totaled 14,000. William George Snodgrass, "A History of the Oklahoma State Federation of Labor to 1918" (master's thesis, University of Oklahoma, 1960), 76; Scales and Goble, *Oklahoma Politics*, 11; J. Green, *Grass-Roots Socialism*, 187–88.

18. Foner, *IWW History*, 515.

19. C. C. Zeigler, the OSFL president, did not accuse Waller of being a Wobbly, but he did claim Waller was in the pay of the National Association of Manufacturers, an accusation also made of the IWW. Waller actually was a former member of the ARU in Omaha before moving to Muskogee and joining the AFL clerks' union. He later lived in a utopian socialist colony at Milton, in Le Flore County; *Unionist*, April 2, 1905; *Oklahoma Labor Unit*, December 16, 23, 1911; May 11, 25, June 1, July 6, December 7, August 31, September 7, November 9, December 28, 1912; September 6, November 19, 1913, April 24, 1914.

"President's Report," *Proceedings of the Ninth Convention of the Oklahoma State Federation of Labor, 1912*, OSFL Collection, WHC/OU, Boxes 70 and 88, 17-18.

20. *Solidarity* became the official IWW newspaper. It was published variously in New Castle, Pennsylvania; Cleveland, Ohio; and Chicago. It also appeared under the names *New Solidarity* and *Industrial Solidarity*, both published in Chicago; "S.," *Industrial Worker*, March 1906; *Solidarity* (Chicago), August 18, 1917; *Daily Oklahoman*, September 23, 1917.

21. "Work of the Labor Lieutenants of Capitalism," *Industrial Worker* (Joliet, Ill.), August 1906.

22. Socialists were influential enough in the OSFL to push through resolutions condemning the state's Democratic Party as "corrupt and rotten" in 1913 and supporting the abolition of the state senate and establishment of a unicameral legislature in 1915. See Snodgrass, "Oklahoma State Federation of Labor," 39, 63, 68; *Official Proceedings of the Twelfth Convention of the Oklahoma State Federation of Labor, 1915*, OSFL Collection, WHC/OU, Boxes 70 and 88, 57; Conlin, *Bread and Roses, Too*, 120-23; Mark Karson, *American Labor Unions and Politics, 1900-1918* (Carbondale: Southern Illinois Univ. Press, 1958), 197.

23. *American Labor Union Journal*, May 13, 1903; "Oklahoma Socialists on Deck," one-page broadsheet, dated Newkirk, January 27, 1900, in Thomas C. Pardo, ed., *Socialist Collections in the Tamiment Library, New York University, 1872-1956* !New York: Microfilming Corporation of America, 1979), Reel 4, f. 30; *Industrial Democrat*, January 1, 1910; *Oklahoma Pioneer*, January 19, March 16, 1910, May 25, 1912; Mrs. Blanche Riehn, "Dear Comrades," dated Muskogee, October 10, 1914, one-page flyer in *Socialist Party of America Papers, 1897-1964* (Glen Rock, N.J.: Microfilming Corporation of America, 1975), Reel 110, Section K; *Tulsa World*, October 9, 1914; News and Views, *International Socialist Review* 17 (January 1917): 443; ibid. (March 1917): 573.

24. For examples, see *New Century*, January 24, 1913; *Beckham County Advocate*, April 24, May 8, 15, 29, September 4, 25, 1913; Daniel Gibson to editor, *Social Democrat*, May 7, 1913. Other examples of SP newspaper articles sympathetic to the IWW include *Industrial Democrat*, February 5, 1910, which reprinted an article on direct action by Robert Hunter, and the *Grant County Socialist*, September 14, 21, 28, October 19, December 14, 1912; *Oklahoma Pioneer*, September 10, 1910; News and Views, *International Socialist Review* 18 (December 1917): 315-16; ibid. (January 1918): 378; Jasper Roberts, "William Haywood" and "Radicals," *Beckham County Advocate*, May 15, 1913.

25. Foote became acting editor of the Spokane-based version of the *Industrial Worker* (not connected with the earlier publication) after the regular editor was arrested during the Spokane free-speech fight of November 1909. Foote too was arrested and sentenced to six months in county jail on a conspiracy charge. See Elizabeth Gurley Flynn, "The Free-Speech Fight at Spokane" in Philip S. Foner, ed., *Fellow Workers and Friends: IWW Free-Speech Fights as Told by the Participants* (Westport, Conn.: Greenwood Press, 1981), 36-45; J. Green, *Grass-Roots Socialism*, 13-14, 20, 27; *Medford Challenge*, August 1, 29, 1904; E. J. Foote, "Break Ranks! Come with Us!" *Industrial Worker* (Joliet, Ill.), June 1906, and "A Voice from the Ranks," ibid., August 1906; *Industrial Worker* (Spokane), June 10, 1909.

26. The *Indian Journal*, April 1, 1886, describes how Knights coal miners in Savannah, in Indian Territory, attempted to kill an engine by draining its boilers and smashing the connecting rods. Conlin, *Bread and Roses, Too*, 104-107; *Guthrie Daily Leader*, July 13, 1894; *Daily Oklahoma State Capital*, June 7, 1894; Almont Lindsey, *The Pullman Strike* (Chicago: Univ. of Chicago Press, 1942), 258; J. Thompson, *Closing the Frontier*, 60-61.

27. For the Wobblies, sabotage specifically emphasized passive resistance. Discussion of sabotage can be found in Thorstein Veblen, "On the Nature and Uses of Sabotage," *Dial* 66 (April 5, 1919): 341-42; Stanley B. Mathewson, *Restriction of Output among Unorganized Workers* (New York: Viking Press, 1931), 6, 6 n; Grover H. Perry, "Sabotage," *Solidarity*, February 1, 1913; ibid., June 4, 1910, February 25, 1911.

28. Mathewson attributed the use of sabotage and other tactics by nonunion workers to speed-ups by management, to management's concern with output and not with workers' input, and to the failure of scientific management to instill confidence in workers. Because piecerates meant that pay-per-piece dropped the harder employees worked, the workers believed they were penalized rather than rewarded for good work; Mathewson, *Restriction of Output*, 146-47; Walker C. Smith, "Sabotage: Its History, Philosophy, and Function" (Spokane: Smith, 1913); *Eleventh Special Report of the U.S. Commission of Labor* (Washington, D.C.: GPO, 1904), 29.

29. In a February 20, 1913, editorial, *Industrial Worker* (Seattle) called Ameringer the "court fool for King Berger" because he attacked sabotage, especially the placing of soap in scab-produced beer, as had happened in a recent Milwaukee brewery strike; *Oklahoma Pioneer*, May 25, 1912; Kipnis, *American Socialist Movement*, 391; Shannon, *Socialist Party of America*, 75.

30. Ameringer and Texas SP member Tom Hickey tried to find a compromise at the 1912 convention, which endorsed industrial unionism but not the IWW. Haywood supported the compromise, which also endorsed political action, but the measure failed to gain support after the reformers introduced the antisabotage amendment. Interestingly enough, the women members of the SP voted solidly for the amendment. Most female party delegates opposed the IWW because the Wobblies believed the ballot was useless for working women. See Bruce Dancis, "Socialism and Women in the United States, 1900-1917," *Socialist Revolution* 6 (January-March 1976): 124; J. Green, *Grass-Roots Socialism*, 217-18; Brissenden, *The IWW*, 280-81; Kipnis, *American Socialist Movement*, 403, 408

31. The left wing's attempts to "recall the recall" and to repeal the antisabotage amendment both failed; Foner, *IWW History*, 394-96, 408-10; Kipnis, *American Socialist Movement*, 380-86, 411-17; Shannon, *Socialist Party of America*, 77-78; J. Green, *Grass-Roots Socialism*, 278-79; "John D.," "Haywood's Recall," *Solidarity*, March 22, 1913; *Social Democrat*, January 15, 1913; *Grant County Socialist*, January 25, 1913; "Socialist Party of America Weekly Bulletin" (Chicago), March 1, 1913, in *Socialist Party Papers*, Reel 4.

32. SP membership dropped after the passage of the antisabotage amendment, from 150,000 to 118,000. Another 40,000 left after Haywood's recall. Kipnis, *American Socialist Movement*, 403, 418; Karson, *American Labor Unions*, 197; William Preston, Jr., *Aliens and Dissenters: Federal Suppression of Radicals, 1903-1933* (Cambridge: Harvard Univ. Press, 1963), 50.

33. As a result of this split, the WFM began to distance itself from the IWW, refusing to pay dues to either faction and finally seceding completely in 1907. Also in 1907, the Trautmann-DeLeon-St. John faction started a new newspaper, the *Industrial Union Bulletin*, which in its March 2, 1907, issue called the Sherman schism "a housecleaning" that would allow the IWW to move ahead. Foner, *IWW History*, 71-75, 78; Dubofsky, *We Shall Be All*, 107, 110-19; *Industrial Worker* (Joliet, Ill.), January 1907, March 1907.

34. *Industrial Union Bulletin*, April 20, 1907; E. J. Foote, Wichita, Kansas, to W. E. Trautmann, ibid., September 29, 1907; Trautmann to Foote, ibid., November 2, 1907.

35. Trautmann to Foote, ibid., November 2, 1907. On Forberg's background, see ibid., May 11, 1907, and *Industrial Worker* (Joliet, Ill.) January 1907.

36. The DeLeon split is detailed in Dubofsky, *We Shall Be All*, 133-41; "M. H.," Lenapah, Okla., December 22, 1919, to editor, *Industrial Union News* (Detroit: WIIU), January 3, 1920.

37. Brissenden, *The IWW*, 177; *Fourth Annual Report of the Oklahoma State Department of Labor, 1911* (Oklahoma City: Oklahoma Department of Labor, 1911), 9; Asfahl, "Oklahoma and Organized Labor," 130-31.

38. No biography exists of Little, despite what appears to be considerable newspaper and archival material. For Little's career in the WFM, see Vernon H. Jenson, *Heritage of Conflict: Labor Relations in the Nonferrous Metals Industry* (Ithaca, N.Y.: Cornell Univ. Press, 1950), 360-61; Philip Mellinger, "'The Men Have All Become Organizers': Labor Conflict and Unionization in the Mexican Mining Communities of Arizona, 1900-1915," *Western Historical Quarterly* 23 (August 1992): 334; Dubofsky, *We Shall Be All*, 118, 135, 186; *Oklahoma Federal Tract Books*, vol. 7, p. 137, microfilm copy in Library, Oklahoma Historical Society, Oklahoma City; obituary of W. R. Little, *Stillwater Gazette*, January 19, 1899; Barbara Lewis, Helena, Montana, July 1, 1983, to Fred Thompson, in Fred Thompson Collection, Box 14, Folder 11, Archives of Labor and Urban Affairs, Walter P. Reuther Library, Wayne State University, Detroit, Michigan (hereafter: ALUA/WSU). Lewis is a third cousin of Little, while Thompson, a member of the IWW since 1920, wrote the union's official history.

39. Little's background in the free-speech fights is covered in Dubofsky, *We Shall Be All*, chap. 8, as well as in Foner, ed., *Fellow Workers and Friends IWW Free-Speech Fights as Told by the Participants*, chaps. 2, 3, and 8. Articles and monographs on individual free-speech fights are also common. *Industrial Worker* (Seattle), April 30, 1910, January 5, 1911; Brad Agnew, "Wagoner, I.T.: 'Queen City of the Prairies,'" *Chronicles of Oklahoma* 64 (winter 1986-87): 42.

40. Frank Little, Guthrie, Okla., November 23, 1911, to editor, *Solidarity*, December 2, 1911, June 14, 1914; *Industrial Worker* (Seattle), February 22, December 26, 1912; Covington Hall, "Labor Struggles in the Deep South" (unpublished manuscript, photocopy in the Archives of Labor and Urban Affairs, Walter P. Reuther Library, Wayne State University, Detroit), 218-19.

41. Little to *Solidarity*, ibid.; Hall, ibid.

42. Helen Marot, *American Labor Unions* (New York: Henry Holt, 1914), 264; *Industrial Worker* (Seattle), January 15, 1910.

Chapter 2. Harvesting the Harvesters

1. Robert F. Hoxie attended the 1913 IWW convention and estimated membership based on paid-up receipts. See Hoxie, "The Industrial Workers of the World and Revolutionary Unionism," in *Readings in Labor*

Economics and Labor Relations, rev. ed., ed. Richard L. Rowan (Homewood, Ill.: Richard L. Irwin, Inc., 1972), 129-31; Brissenden, *The IWW*, 336-37; *Solidarity*, August 2, 22, November 2, 1913, February 6, March 7, 14, 21, 1914, May 15, 1915; Frank S. Hamilton, "A Screed and a Suggestion," *Solidarity*, November 21, 1914.

2. Eric H. Monkkonen, "Introduction," in *Walking to Work: Tramps in America, 1790-1935*, ed. Eric H. Monkkonen (Lincoln: Univ. of Nebraska Press, 1984), 3; David E. Schob, *Hired Hands and Plowboys: Farm Labor in the Midwest, 1815-1860* (Urbana: Univ. of Illinois Press, 1975), 81, 91, 107, 109, 110.

3. Jesse Harder, "Wheat Production in Northwestern Oklahoma, 1893-1932" (master's thesis, University of Oklahoma, 1952), 13; Allen Gale Applen, "Migratory Harvest Labor in the Midwestern Wheat Belt, 1870-1940" (Ph.D. dissertation, Kansas State University, 1974), 27, 71-72; Douglas Hale, "The People of Oklahoma: Economic and Social Change" in *Oklahoma: New Views of the Forty-Sixth State*, ed. Anne Hodges Morgan and H. Wayne Morgan (Norman: Univ. of Oklahoma Press, 1982), 46-47, 56-57; Peeler, *Standard Blue Book*, 137-38; Isaac F. Marcosson, "Harvesting the Wheat," *World's Work* 9 (November 1904): 54-59.

4. Large farms were growing in Oklahoma as early settlers were bought out by newer, wealthier immigrants. Many of these farms undoubtedly were set up like the bonanza farms of the Red River Valley of the North, where numerous professional management and business practices were adapted to agriculture. Such practices were already being applied to cotton farms in southeastern Oklahoma. Applen, "Migratory Harvest Labor," 122; Ellsworth Collings and Alma Miller England, *The 101 Ranch* (Norman: Univ. of Oklahoma Press, 1971), 87-88; I. K. Friedman, "Our Inland Migrations," *World's Work* 8 (July 1904): 5287-88, 5290; Hiram Drache, *Day of the Bonanza* (Fargo: North Dakota Institute of Regional Studies, 1964), 5; Don D. Lescohier, *Conditions Affecting the Demand for Harvest Labor in the Wheat Belt*, Department of Agriculture Bulletin no. 1030 (Washington, D.C.: GPO, 1924), 5-14; Don D. Lescohier, "Hands and Tools of the Wheat Harvest," *Survey* 50 (July 1, 1923): 378; Thomas D. Isern, *Bull Threshers and Bindlestiffs: Harvesting and Threshing on the North American Plains* (Lawrence: Univ. Press of Kansas, 1990), 138-39; *El Reno News*, July 16, 1897; *Stillwater Advance*, June 20, 1901; *New Century*, January 31, 1913; *Solidarity*, May 10, 1913.

5. Kansas Agricultural College estimated the need for harvest labor at one man for each fifty acres, but Lescohier found the number was closer to one for each one hundred acres. Isern, *Bull Threshers*, 138-39;

Custom Combining on the Great Plains: A History (Norman: Univ. of Oklahoma Press, 1981), 5; Don D. Lescohier, *Harvest Labor Problems in the Wheat Belt*, Department of Agriculture Bulletin no. 1020 (Washington, D.C.: GPO, 1922), 1; J. H. Arnold and R. R. Spafford, "Farm Practices in Growing Wheat," in *Yearbook of the United States Department of Agriculture, 1919* (Washington, D.C.: GPO, 1920), 137-38, 143-45.

6. Applen, "Migratory Harvest Labor," 74; Lescohier, *Demand for Harvest Labor*, 6-14; Isern, *Custom Combining*, 5-8; Isern, *Bull Threshers*, 26-27, 77-78; Arnold and Spafford, "Farm Practices," 143-45; Harder, "Wheat Production," 28-29.

7. Wobbly Joseph Murphy said the hi-jack's practice of throwing victims from trains was called "greasing the rails." See "Joseph Murphy Interview" in Stewart Bird, Dan Georgakas, and Deborah Shaffer, eds., *Solidarity Forever: An Oral History of the IWW* (Chicago: Lake View Press, 1985), 46; Monkkonen, "Introduction," 9-11; James Forbes, "The Tramp, or Caste in the Jungle," *Outlook* 98 (August 19, 1911): 871; *Daily Oklahoman*, May 23, 1915.

8. Ralph Chaplin, *Wobbly: The Rough-and-Tumble Story of an American Radical* (Chicago: Univ. of Chicago Press, 1949), 87-88. See also *Solidarity*, February 13, 1915.

9. Carey McWilliams, *Ill Fares the Land: Migrants and Migratory Labor in the United States* (Boston: Little, Brown, 1942), 94-95; Lescohier, "Hands and Tools," 378-79; *Solidarity*, August 1, 1914.

10. Schob, *Hired Hands and Plowboys*, 94-96; "Jack Miller Interview," in Bird, Georgakas, and Shaffer, *Solidarity Forever*, 37.

11. Isern, *Bull Threshers*, 103-104; *Daily Oklahoman*, May 23, 1915.

12. "Joseph Murphy Interview," in Bird, Georgakas, and Shaffer, *Solidarity Forever*, 42-43; Isern, *Bull Threshers*, 163-64; Lescohier, "Hands and Tools," 410; "Testimony of William Casebolt" in Harris George, *The IWW Trial* (Chicago: Industrial Workers of the World, 1919), 93-94; "The Happy Life of the Harvest Hand," *Solidarity*, August 1, 1914.

13. Isern argues that only a few disreputable or financially strapped farmers would not pay and that the need for steady labor over the years generally held such behavior in check. He does admit, however, the bindlestiffs lacked any legal recourse to recover lost wages. Isern, *Bull Threshers*, 154; "Joseph Murphy Interview," in Bird, Georgakas, and Shaffer, *Solidarity Forever*; "Nels Peterson Interview," ibid., 116; Tom Connors, "The Industrial Union in Agriculture," in *Twenty-Five Years of Industrial Unionism* (Chicago: Industrial Workers of the World, 1930), 36. copy in Labadie Collection, University of Michigan, Ann Arbor.

14. *Solidarity,* July 10, 1915; *Industrial Worker* (Seattle), January 16, 23, 30, 1913; Ethel E. Carpenter, "The Farmer and His Hired Hand," *Industrial Worker* (Joliet, Ill.), 1 (April 1906).

15. Nels Anderson's use of the terms matches that of the IWW itself. Paul T. Ringenbach notes the term "tramp" was in common usage by the 1890s and became interchangeable with "hobo" after the Panic of 1907. In a Works Progress Administration study, John N. Webb noted it was also popular habit to call the migratories "workers" when the grain needed harvesting and "bums" during the slack season. Nels Anderson, *The Hobo: The Sociology of the Homeless Man* (Chicago: University of Chicago, 1961), 87, 89-95, 96-98; Paul T. Ringenbach, *Tramps and Reformers, 1873-1916: The Discovery of Unemployment in New York* (Westport, Conn.: Greenwood Press, 1973), 18, 25-26, 191; John N. Webb, *The Migratory-Casual Worker,* Works Progress Administration, Division of Social Research, Research Monograph no. 7 (Washington, D.C.: GPO, 1937), 2; Charles Ely Adams, "The Real Hobo: What He Is and How He Lives," *Forum* 33 (June 1902): 439; O. F. Lewis, "The Tramp Problem," *Annals of the American Academy of Political and Social Sciences* 40 (March 1912): 217; Don D. Lescohier, "Harvesters and Hoboes in the Wheat Fields," *Survey* 50 (August 1, 1923): 482-83; John C. Schneider, "Tramping Workers, 1890-1920: A Subcultural View," in *Walking to Work: Tramps in America, 1790-1935,* ed. Eric H. Monkkonen (Lincoln: Univ. of Nebraska Press, 1984), 217-18; *Industrial Pioneer,* 2d ser., 1 (October 1923): 10.

16. The term "sons of rest" came from a song by Nat W. Wills, a turn-of-the-century vaudeville comedian who created the image of the "happy tramp." The song itself implies a familiarity with whiskey, reinforcing another stereotype. Ringenbach, *Tramps and Reformers,* xiv-v; *Daily Oklahoman,* June 21, 1914; *Enid Daily Eagle,* May 30, 1915, June 1, 12, 1916, June 5, 1917; *Tulsa World,* June 3, 1915.

17. John J. McCook counted only thirteen African-Americans in his total of 1,349 tramps, or less than 1 percent. None was from the South. Regarding Mexican migrants, Frederick M. Noa's contemporary account said many Mexican railroad workers joined the wheat and cotton harvests for better pay. John J. McCook, "A Tramp Census and Its Revelations," *Forum* 15 (August 1893): 756-57; Frederick M. Noa, "Mexican Peons in the U.S.," *Oklahoma Pioneer,* January 28, 1911; Anderson, *The Hobo,* 107-108, 150; Schneider, "Tramping Workers," 218-19; Forbes, "The Tramp," 874-75; Ringenbach, *Tramps and Reformers,* 71-72; McWilliams, *Ill Fares the Land,* 92-93; Nils H. Hanson, "Among the Harvesters," *International Socialist Review* 16 (August 1915): 76-77;

Eric H. Monkkonen, "Regional Dimensions of Tramping, North and South: 1880-1910," in *Walking to Work: Tramps in America, 1790-1935*, ed. Eric H. Monkkonen (Lincoln: Univ. of Nebraska Press, 1984), 204.

18. Only 40 of the 1,349 tramps McCook interviewed blamed alcohol for their condition; 16 said they had a roving disposition; while 82.8 percent said they were actively seeking work. McCook said 57.4 percent of the tramps he talked to had had trades or skilled jobs, while 41.4 were unskilled. The percentage of tramps who were unskilled was probably higher by 1914. Lescohier, in his study of the 1920 harvest, noted several industrial workers, including a miner, a carpenter, a glass blower, and an embalmer. College students were rare in the harvest crews, but farmers apparently considered them good workers. Lescohier counted about 800 in the 1920 harvest, roughly 300 of them at Woodward. Mike Davis contends many tramps were skilled workers who took to the road because American unions, unlike British ones, did not provide stationary out-of-work benefits and there was no state-supported unemployment relief system. McCook, "A Tramp Census," 754-55; Lescohier, "Hands and Tools," 376, 379-82; Mike Davis, "Forced to Tramp" in *Walking to Work: Tramps in America, 1790-1935*, ed. Eric H. Monkkonen (Lincoln: Univ. of Nebraska Press, 1984), 155; Schneider, "Tramping Workers," 213-15; McWilliams, *Ill Fares the Land*, 92-93; Applen, "Migratory Harvest Labor," 77, 90-91; *Daily Oklahoman*, June 21, 1914.

19. Frederick Ryan has noted that the IWW had an agreement to recognize UMW members during the harvest and let them ride the trains. Regarding the term "scissorbill," Anderson suggested it originally meant a man who carried tools to sharpen saws, knives, and the like. For the IWW, it meant a worker who was not class-conscious and who sided with the boss. Frederick L. Ryan, *Problems of the Oklahoma Labor Market, with Special Reference to Unemployment Compensation* (Oklahoma City: Semco Color Press, 1937), 27; Anderson, *The Hobo*, 99; Monkkonen, "Afterword," in *Walking to Work: Tramps in America, 1790-1935*, ed. Eric H. Monkkonen (Lincoln: Univ. of Nebraska Press, 1984), 238; Lescohier, "Harvesters and Hoboes," 482-83; Webb, *Migratory-Casual Worker*, xviii, 17.

20. *Proceedings of the Sixth Annual OSFL Convention, 1909*, OSFL Collection, WHC/OU, Boxes 70 and 88, 23, 41; *Proceedings of the Seventh Annual OSFL, 1910*, ibid., 91; *Proceedings of the Eighth Annual OSFL Convention, 1911*, ibid., 93, 105.

21. Anderson lists several causes for hoboing, including unemployment or physical disability. But Anderson also includes what he calls "personality defects," including alcoholism and homosexuality. He

contends hoboes likely turned to homosexuality because of sexual isolation from women on the road. Schneider notes the relationships were usually between adult males, nicknamed "gay-cats" or "wolves," and runaway boys, but Forbes discusses the kidnapping of boys, suggesting a more exploitative relationship. Peter Stearns suggests that working class culture as a whole was intolerant of homosexuality as it challenged the values essential to masculinity in a class regularly subjected to the authority of others. Anderson, *The Hobo*, 85-86, 144-49; Schneider, "Tramping Workers," 213-15; Forbes, "The Tramp," 870-71, 873; Peter N. Stearns, *Be a Man! Males in Modern Society* (New York: Holmes and Meier Publishers, 1979), 61.

22. A good discussion of the migratories' bachelor culture is found in Peter Way's work on nineteenth-century canal laborers, much of which is also valid for the early twentieth century. In a similar vein, Peter Stearns has suggested that the preindustrial notion that a propertyless man was not fully a man survived into the industrial era. The unskilled, lacking even a trade to define their manliness, asserted their masculinity through drinking, brawling, and whoring. In addition, Peter Filene has noted that Victorian society saw women and the family as a bulwark against male lust. An all-male society, such as that of the migratories, was therefore uncontrolled and could tend only toward vice, corruption, and violence. Peter Way, "Evil Humors and Ardent Spirits: The Rough Culture of Canal Construction Laborers," *Journal of American History* 79 (March 1993): 1398-1401; Stearns, *Be a Man!* 44-45, 63, 66, 80-83; Peter Filene, *Him/Her/Self: Sex Roles in Modern Society*, rev. ed., (Baltimore: Johns Hopkins Univ. Press, 1992), 84-85; Jon M. Kingsdale, "'The Poor Man's Club': Social Functions of the Urban Working Class Saloon," in *The American Male*, ed. Elizabeth H. Pleck and Joseph H. Pleck (Englewood Cliffs, N.J.: Prentice Hall, 1980), 258-59, 261-62, 267; Jimmie Lewis Franklin, *Born Sober: Prohibition in Oklahoma, 1907-1959* (Norman: Univ. of Oklahoma Press, 1971), xii, 3, 5-6, 36-38; Anderson, *The Hobo*, 144-49, 249; *Enid Daily Eagle*, June 17, 1915.

23. Ringenbach notes that as early as 1877 some Americans saw tramps as the "shock troops" of violent revolution. Ringenbach, *Tramps and Reformers*, 13; Davis, "Forced to Tramp," 142; *Tulsa World*, June 3, 1915.

24. *Enid Daily Eagle*, June 23, 1915; Miller, *Oklahoma Populism*, 91-93, 120.

25. Connors, "Industrial Union in Agriculture," 36-37.

26. The term "I Won't Work" probably came from agitational stickers that read "I Won't Work more than 8 Hours per Day," which appeared in the Wheat Belt by 1912. See E. W. Latchem, "Those Terrible I Won't

Notes to Pages 45-49   215

Works," undated one-page typescript in Box 145, Folder 29, Industrial Workers of the World Collection, Archive of Labor and Urban Affairs, Walter P. Reuther Library, Wayne State University, Detroit, Michigan (hereafter IWW Collection, ALUA/WSU); "Joseph Murphy Interview," in Bird, Georgakas, and Shaffer, *Solidarity Forever*, 46; *Solidarity*, February 13, July 10, 1915.

27. Way, "Evil Humors," 1398-1401; Stearns, *Be a Man!* 63, 66.

28. Applen, "Migratory Harvest Labor," 150; Stuart Jamieson, *Labor Unionism in American Agriculture*, Department of Labor Bulletin no. 836 (Washington, D.C.: GPO, 1945), 398; *American Labor Union Journal*, June 11, 1903; *Labor World*, February 14, 1904; *Solidarity*, April 8, 1911, August 1, 1914; Carpenter, "Farmer and Hired Hand"; Foner, *IWW History*, 77.

29. Brissenden, *The IWW*, 333-34; Fred Thompson and Patrick Murfin, *The IWW: Its First Seventy Years* (Chicago: Industrial Workers of the World, 1976), 93

30. Applen, "Migratory Harvest Labor," 142-44, 153; Udo Sautter, *Three Cheers for the Unemployed: Government and Unemployment before the New Deal* (New York: Cambridge Univ. Press, 1991), 71, 75n; O. L. Harvey, ed., *The Anvil and the Plow* (Washington, D.C.: Department of Labor/GPO, 1963), 17; *Vinita Weekly Chieftain*, March 4, 1910.

31. Dubofsky, *We Shall Be All*, 294-96; Foner, *IWW History*, 476.

32. George Creel, "Harvesting the Harvest Hands," *Harper's Weekly*, September 26, 1914, 292.

33. Ibid., 292-93; *Daily Oklahoman*, May 5, June 11, 12, 1914; *Oklahoma Labor Unit*, June 13, 27, 1914.

34. Creel, "Harvesting," 292-93; *Daily Oklahoman*, June 10, 17, 21, 1914.

35. Creel, "Harvesting," 292-93; *Daily Oklahoman*, June 22, 1914; *Solidarity*, June 13, 20, 1914; *Oklahoma Labor Unit*, June 6, 1914.

36. *Solidarity*, August 22, October 3, 10, November 28, 1914; February 20, April 24, 1915; September 23, 1916; Foner, *IWW History*, 474-75; Brissenden, *The IWW*, 330.

37. F. Thompson and Murfin, *First Seventy Years*; Dubofsky, *We Shall Be All*, 314-15; *Solidarity*, June 26, 1915; E. W. Latchem, "Some Vitally Important Background Information [comments on the AWO minutes]," three-page typescript dated October 8, 1965, 3, copy in Box 75, Folder 1, E. W. Latchem Papers, Small Processed Collections, ALUA/WSU.

38. Latchem, "Some Vitally Important Background," 2; E. W. Latchem, "To the Harvest Workers," *Solidarity*, February 27, 1915; E. W. Latchem, South Dakota, to editor, *Solidarity*, October 9, 1915; *Solidarity*, February 20, 1915.

39. Ibid., February 13, May 1, June 5, 1915.

40. Ibid., June 12, 1915; F. Thompson and Murfin, *First Seventy Years*, 94; E. Workman, "History of the 400, AWO, the One Big Union Idea in Action" (New York: One Big Union Club, 1939), 10, copy of pamphlet in Box 163, IWW Collection, ALUA/WSU.

41. Jamieson, *Labor Unionism*, 400; Workman, "History of the 400," 11; Sautter, *Three Cheers*, 85-86.

42. Workman, "History of the 400"; *Solidarity*, June 26, July 3, 17, 1915.

43. *Enid Daily Eagle*, May 26, 27, 31, 1915.

44. *Enid Daily Eagle*, May 31, June 1, 1915; *Daily Oklahoman*, June 1, 1915; *Tulsa World*, June 1, 1915.

45. *Enid Daily Eagle*, June 7, 15, 23, 1915; *Solidarity*, June 26, July 31, 1915; Hanson, "Among the Harvesters," 76; Jamieson, *Labor Unionism*, 400-401; Workman, "History of the 400."

46. *Solidarity*, June 26, September 4, 1915.

47. Foner, *IWW History*, 477; *Industrial Worker*, May 27, 1916.

48. Foner, *IWW History*, 478; *Industrial Worker*, June 10, 1916; *Solidarity*, November 27, 1915; *Oklahoma News*, February 9, 1916; *Daily Oklahoman*, June 12, 1916.

49. *Solidarity*, June 10, 17, 24, 1916; *Daily Oklahoman*, June 12, 1916.

50. Chaplin, *Wobbly*, 207-208; Foner, *IWW History*, 478-79, 483; *Solidarity*, June 24, 1916; *Daily Oklahoman*, June 12, 1916.

51. In a letter to Ben Williams, the editor of *Solidarity*, dated January 13, 1916, Bill Haywood cited cases of hold-up men separating IWW members from nonmembers and then robbing only the nonunion men. Quoted in *Evidence and Cross-Examination of William D. Haywood in the Case of the United States v. William D. Haywood, et al.* (Chicago: Industrial Workers of the World, 1918), 175; Chaplin, *Wobbly*; Foner, *IWW History*, 476-77, 482; "Joseph Murphy Interview," in Bird, Georgakas, and Shaffer, *Solidarity Forever*, 43-46; Thompson and Murfin, *First Seventy Years*, 94-95; *Industrial Worker*, August 26, 1916.

52. *Evidence and Cross-Examination of William D. Haywood*, 182-83; "Jack Miller Interview," in Bird, Georgakas, and Shaffer, *Solidarity Forever*, 39; Foner, *IWW History*, 478-79, 484; F. Thompson and Murfin, *First Seventy Years*, 94-95.

53. Philip Taft notes that some criminals joined the IWW as a cover for their activities. See Taft, "The IWW in the Grain Belt," *Labor History* 1 (winter 1960): 62; Brissenden, *The IWW*, 39; Dubofsky, *We Shall Be All*, 317; report from E. N. Osborne in Mitchell, South Dakota, to editor, *Solidarity*, September 6, 1916; *Solidarity*, November 11, 18, 1916.

54. *Solidarity*, July 13, 1914; Forrest Edwards, "The Class War in Harvest Country," *Solidarity*, August 19, 1916.

Chapter 3. Organizing "Oily Willy"

1. James S. Koen to John L. Burk, Winslow, Arkansas, August 10, 1916, carbon copy in Labor Union folder, vertical files, Library, Oklahoma Historical Society, Oklahoma City. Koen to William D. Haywood, Chicago, August 1916, quoted in *Evidence and Cross-Examination of William D. Haywood*, 180-81.

2. Taft, "IWW in the Grain Belt," 56.

3. C. W. Shannon, *Handbook on the Natural Resources of Oklahoma* (Norman: Oklahoma Geological Survey, 1916), 15; Gibson, *Oklahoma: A History*, 270-72; Steffen, "Stages of Development," 26-27; J. Thompson, *Closing the Frontier*, 117-18; Carl Coke Rister, *Oil! Titan of the Southwest* (Norman: Univ. of Oklahoma Press, 1949), 89-92, 119-22, 125; Carl Solberg, *Oil Power: The Rise and Imminent Fall of an American Empire* (New York: Mentor Books, 1976), 61-62.

4. Steffen, "Stages of Development," 26-27; Rister, *Oil!*, 137-42; Solberg, *Oil Power*, 55-56; John G. McLean and Robert William Haigh, *The Growth of Integrated Oil Companies* (Boston: Harvard University Graduate School of Business Administration, 1954), 256-57; *Solidarity*, August 29, 1914.

5. Solberg, *Oil Power*, 62-65; McLean and Haigh, *Growth of Integrated Oil*, 255.

6. Scales and Goble, *Oklahoma Politics*, 9-10, 30; *Daily Oklahoman*, September 11, 1907; *Industrial Democrat*, September 17, 1910; Albert Raymond Parker, "Life and Labor in the Mid-Continent Oil Fields" (Ph.D. dissertation, University of Oklahoma, 1951), 122; Upton Sinclair, *The Brass Check: A Study of American Journalism*, rev. ed. (Pasadena: Sinclair, 1931), 241.

7. Several contemporary observers commented on the level of vice in oil towns. E. P. Hill, a McAlester attorney, estimated one hundred places in Tulsa sold illegal liquor, while Lt. Paul Popenoe estimated twenty-two Tulsa hotels harbored prostitutes. He added that military records showed nearly 43,000 Oklahoma recruits had either syphilis or gonorrhea, a number that seems somewhat exaggerated. E. P. Hill, Tulsa, to Oklahoma Attorney General S. Freeling, October 26, 1918; Lt. Paul Popenoe, Tulsa, to Freeling, May 12, 1918; Mrs. Phil Hill, Oilton, to Freeling, December 6, 1918; and B. F. Grigsby and Oliver M. Ingenthron, Davenport, to Freeling, January 5, 1919. All letters in Box 19, Folder 940, State

Attorney General's Files, Oklahoma State Archives Department of Libraries, Oklahoma City (hereafter AG Files, OSA); Steffen, "Stages of Development," 28; Parker, "Life and Labor," 46-47; C. B. Glasscock, *Then Came Oil: The Story of the Last Frontier* (Westport, Conn.: Greenwood Press, 1976), 277; Drue Lemuel Deberry, "The Ethos of the Oklahoma Oil Boom Frontier" (master's thesis, University of Oklahoma, 1970), 20; Roger Olien and Diana Davids Olien, *Oil Boom: Social Change in Five Texas Towns* (Lincoln: Univ. of Nebraska Press, 1982), 128-31; George Carney, "The Historic Preservation of the Cushing Oil Field: A Summary," *Payne County Historical Review* 1 (April 1981): 7.

8. Parker, "Life and Labor," 76; *Oil and Gas Journal* (Tulsa), September 20, 1917, 3; *Harlow's Weekly*, September 25, 1915, 253; James Leslie Gilbert, "Three Sands: Oklahoma Oil Field and Community of the 1920s" (master's thesis, University of Oklahoma, 1967), 95-96.

9. Parker, "Life and Labor," 76; D. Earl Newsom, *Drumright II: A Thousand Memories* (Perkins, Okla.: Evans Publications, 1987), 16-17.

10. Some hotels were even mobile, moving from boom town to boom town. Olien and Olien, *Oil Boom*, 45-46; Newsom, *Drumright II*, 16-17.

11. Olien and Olien, *Oil Boom*, 22-26; Hale, "People of Oklahoma," 59; "In the Oil Fields," *International Socialist Review* 14 (May 1914), 666; Webb, *Migratory-Casual Worker*, 6-7.

12. The roustabouts were nicknamed "hideabouts" from an older farm expression that "you never see hide nor hair of them till supper." Parker, "Life and Labor," 47; Olien and Olien, *Oil Boom*, 6, 29; Deberry, "Ethos of the Frontier"; "Organizing the Oil Industry, No. 4: The Lease Workers," *Solidarity* (Chicago), July 30, 1921.

13. Lescohier, "Harvesters and Hoboes," 486; James Foy, "A Migratory Worker's Diary," *Industrial Pioneer* 1 (February 1924): 29; "The Old Hand" in Box 43, Folder 6, Ned DeWitt WPA Writers Project on Oil in Oklahoma, Western History Collections, University of Oklahoma, Norman (hereafter: Oil in Oklahoma, WHC/OU).

14. "In the Oil Fields," 664-66; Olien and Olien, *Oil Boom*, 100; H. C. Fowler, *Accidents in the Petroleum Industry of Oklahoma, 1915-1924*, Department of Commerce Technical Paper no. 392 (Washington, D.C.: GPO, 1926), 14-16.

15. "In the Oil Fields," 666; Earl Bruce White, "The Wichita Indictments and Trial of the Industrial Workers of the World, 1917-1919, and the Aftermath" (Ph.d. dissertation, University of Colorado, 1980), 38; Ryan, *Problems of the Oklahoma Labor Market*, 30; "The Roughneck, No. 1" in Box 42, Folder 21, Oil in Oklahoma, WHC/OU; Ruth A. Allen,

Chapters in the History of Organized Labor in Texas, Publ. no. 4143 (Austin: University of Texas Bureau of Research in the Social Sciences, 1941), 240.

16. Parker, "Life and Labor," 67; *Solidarity*, July 9, 16, 23, 30, 1921; *Cushing Citizen*, October 15, 1914.

17. The oil companies regularly offered the services of their private police to the American Protective League and the Justice Department. See William T. Lampe, *Tulsa County in the World War* (Tulsa: Tulsa County Historical Society, 1919), 152; Parker, "Life and Labor," 87-88; Allen, *Texas Labor*, 238; "In the Oil Fields," 667.

18. "The Roughneck, No. 1," Oil in Oklahoma, WHC/OU.

19. Harvey O'Connor, *History of the Oil Workers International Union-CIO* (Denver: OWIU-CIO, 1950), 94.

20. Parker, "Life and Labor," 29-30; William Brown interview, Vinita, Oklahoma, March 16, 1938, in Indian-Pioneer Papers, Vol. 12, 233, WHC/OU.

21. "Hot Oil," twenty-five-page typescript in Box 43, Folder 1, Oil in Oklahoma, WHC/OU, 3-4; Olien and Olien, *Oil Boom*, 49: "In the Domain of Standard Oil," *Industrial Pioneer* 1 (September 1923), 17-19.

22. "Hot Oil," Oil in Oklahoma, WHC/OU, 4, 7.

23. "The Rig Builder: In the Mud," eighteen-page typescript in Box 42, Folder 12, Oil in Oklahoma, WHC/OU, 5, 7.

24. Parker, "Life and Labor," 51; *Oil and Gas Journal*, September 6, 1917; Max W. Ball, *This Fascinating Oil Business* (Indianapolis: Bobbs-Merrill, 1940), 178-80, 185-86; George Carney, "Slappin' Collars and Stabbin' Pipe: Occupational Folklife of Old-Time Pipeliners," in *Program Book of Smithsonian Festival of American Folklife, 1982*, (Washington, D.C.: Smithsonian Institution/GPO, 1982), 15-17.

25. Ball, *Fascinating Oil Business*; Carney, "Slappin' Collars," 15-17.

26. Carney, "Slappin' Collars," 15-17; Ball, *Fascinating Oil Business*, 182-85.

27. Newsom, *Drumright II*, 16-17; Parker, "Life and Labor," 34, 120-21.

28. J. M. Healy, Yale, Oklahoma, to editor, February 23, 1915, *Solidarity*, March 6, 1915.

29. Brissenden, *The IWW*, 161, 270-71.

30. Federal Writers Project of Oklahoma, *Labor History*, 69, 101; O'Connor, *OWIU-CIO History*, 4-5; Herbert G. Gutman, "The Labor Policies of the Large Corporation in the Gilded Age: The Case of the Standard Oil Company," in Herbert G. Gutman, *Power and Culture: Essays on the American Working Class*, ed. Ira Berlin (New York: Pantheon Books, 1987), 231-53.

31. Hall, "Labor Struggles in the Deep South," 179; F. Thompson and Murfin, *First Seventy Years*, 68; Fred L. Tiffany, "Appeal to Oil Workers," *Solidarity*, April 26, 1913.

32. *Solidarity*, December 27, 1913; January 31, June 13, July 13, 1914; Jack Allen, "A Review of the Facts Relating to the Free-Speech Fight at Minot, North Dakota, from August 1 to August 17, 1913," and Ed Nolan, "From Frisco to Denver," both in *Fellow Workers and Friends: IWW Free-Speech Fights as Told by Participants*, ed. Philip S. Froner (Westport, Conn.: Greenwood Press, 1981), 152, 158, 160, 162; Parker, "Life and Labor," 64–66.

33. *Solidarity*, February 14, 1914.

34. *Solidarity*, January 31, February 21, March 14, 1914; *Oil and Gas Journal*, April 23, 1914; O'Connor, *OWIU-CIO History*, 94.

35. "In the Oil Fields," 664; *Solidarity*, April 11, 1914; *Oil and Gas Journal*, April 23, 1914.

36. *Oil and Gas Journal*, April 23, 30, 1914.

37. Dubofsky, *We Shall Be All*, 173–74; Preston, *Aliens and Dissenters*, 44; *Daily Oklahoman*, April 27, 1914; *Oil and Gas Journal*, April 23, 30, 1914.

38. *Oklahoma Labor Unit*, June 30, 1914.

39. *Solidarity*, August 29, 1914; Blanche Rien, "Dear Comrades" letter, October 14, 1914, *Socialist Party Papers*, Reel 110.

40. Solberg, *Oil Power*, 66; J. Stanley Clark, *The Oil Century* (Norman: Univ. of Oklahoma Press, 1958), 161–65. (despite some authors' confusion, J. Stanley Clark is unrelated to SP member Stanley J. Clark); J. Thompson, *Closing the Frontier*, 118; *Oil and Gas Journal*, July 2, 1914.

41. *Industrial Worker* (Seattle), December 2, 1916.

42. The IWW union hall in the West functioned as a "home away from home," even providing a kitchen; but western workers engaged more in "hall-room," or idle, talk and less in solid "shop talk" than did eastern workers, who consequently derided the westerners as mere "spittoon philosophers." See Brissenden, *The IWW*, 315–16; *Solidarity*, March 27, May 22, 1915; *Oil and Gas Journal*, April 15, 1915.

43. William D. Haywood, "Report of the General Secretary-Treasurer to the Tenth Convention of the Industrial Workers of the World, November-December 1916," copy in Box 7, Folder 23, IWW Collection, ALUA/WSU, 6, 11; *Oil and Gas Journal*, July 15, 1915; Parker, "Life and Labor," 56.

44. Jack Allen, Oklahoma City, to editor, undated, *Solidarity*, January 15, 1916; *Daily Oklahoman*, December 15, 16, 1915; *Oklahoma Labor Unit*, December 18, 1915.

45. An undated article on the oil industry from the *New York Journal of Commerce*, reprinted in *Solidarity*, noted that, while oil prices first fell because of the war, demand increased once Russian supplies were cut off. The piece estimated that by 1915 Oklahoma led the nation in oil production, with 25,000 barrels a day from the Cushing Field alone. See *Solidarity*, February 12, 1916; J. Thompson, *Closing the Frontier*, 119-20; W. L. Connelly, *The Oil Business as I Saw It: Half a Century with Sinclair* (Norman: Univ. of Oklahoma Press, 1954), 67-68.

46. W. C. King, "Workers: Now Is the Time," *Industrial Worker* (Seattle), June 3, 1916; *Solidarity*, April 10, 1915, January 15, 1916 (quotation is from the January 15 issue).

47. *Drumright Derrick*, April 22, May 3, 1916; *Industrial Worker* (Seattle), December 2, 1916; O'Connor, *OWIU-CIO History*, 94; F. Thompson and Murfin, *First Seventy Years*, 113; "Message to the Oil Workers," undated IWW pamphlet, in Melvyn Dubofsky, ed., *The Strike Files of the U.S. Department of Justice, Part I, 1894-1920* (Bethesda, Md.: University Publications of America, 1990), Reel 8, ff. 201-204; George Shirk, Oklahoma Place-Names, 2d ed. (Norman: Univ. of Oklahoma Press, 1974), 170.

48. "Message to the Oil Workers," in Dubofsky, *Strike Files*, Reel 8, ff. 201-204; *Solidarity*, November 25, 1916.

49. *Solidarity*, November 25, 1916; *Oil and Gas Journal*, November 23, 1916; *Industrial Worker* (Seattle), February 24, 1917; *Drumright Evening News*, February 9, 1917; Joyce Kornbluh, ed., *Rebel Voices: An IWW Anthology* (Ann Arbor: Univ. of Michigan Press, 1964), 87.

50. *Industrial Worker*, February 24, April 7, 1917; *Solidarity*, January 15, 1916, March 24, April 7, 21, 1917; O'Connor, *OWIU-CIO History*, 5, 7; Parker, "Life and Labor," 117.

51. Both Kornbluh and Foner have the name changed to AWIU no. 110 in March 1917, although that change occurred only after the war. Until that time it was known as AWIU no. 400. Kornbluh, *Rebel Voices*, 232; Foner, *IWW History*, 485n.

52. *Solidarity*, April 7, 1917; Earl Bruce White, "The IWW and the Mid-Continent Oil Field," in *Labor in the Southwest: The First One Hundred Years*, ed. James C. Foster (Tucson: Univ. of Arizona Press, 1982), 68-69.

53. White, "IWW and the Mid-Continent," 68-69; report of T. F. Weiss, DOJ agent for Oklahoma City, November 8, 1917, in Randolph Boehm, ed., *United States Military Intelligence Reports: Surveillance of Radicals in the United States, 1917-1941* (Frederick, Md.: University Publications of America, 1984), Reel 6, f. 270 (hereafter *USMIR Microfilm*). This

microfilm series contains information from National Archives Record Group 165, File Series 10110 and 10058.

54. *Tulsa Democrat*, August 11, 1917.

Chapter 4. With Folded Arms? Or with Squirrel Guns?

1. The Green Corn Rebellion has been the subject of at least three works, all heavily flawed and all erroneously tying the IWW directly to the rebellion. The best-known is Charles Bush's "The Green Corn Rebellion," which is badly dated, heavily biased against the rebels, and filled with errors. Those errors are repeated by Sherry Harrod Warrick and John J. Womack, Jr. Warrick's work is also biased and uses convoluted logic to "prove" IWW involvement. Womack, while more sympathetic to the rebels, also assumes the IWW directed the uprising. A new study is clearly needed. Charles Bush, "The Green Corn Rebellion" (master's thesis, University of Oklahoma, 1932); Sherry Harrod Warrick, "Antiwar Reaction in the Southwest during World War I" (master's thesis, University of Oklahoma, 1973); John J. Womack, Jr., "Oklahoma's Green Corn Rebellion: The Importance of Fools" (senior thesis, Yale University, 1961); *Shawnee News Herald*, August 3, 1917; *Wewoka Democrat*, August 9, 1917.

2. Robert Bruere, "The Industrial Workers of the World: An Interpretation," *Harper's Monthly* 137 (September 1918): 256; *Harlow's Weekly*, September 26, 1917, 6; *Daily Oklahoman*, October 21, 1917; *Tulsa World*, November 6, 7, 1917; Sen. Harry F. Ashurst, "The IWW Menace," *Congressional Quarterly* 55 (1917): 6687.

3. William Bennett Bizzell, *Farm Tenantry in the United States* (College Station, Texas: Texas Agricultural Experiment Station, 1921), 159, 161; William J. Spillman and E. A. Goldenweiser, "Farm Tenantry in the United States," in *Yearbook of the United States Department of Agriculture, 1916* (Washington, D.C.: GPO, 1917), 321-22.

4. Angie Debo, *The Rise and Fall of the Choctaw Republic*, 2d ed. (Norman: Univ. of Oklahoma Press, 1961), 111; Spillman and Goldenweiser, "Farm Tenantry," 321-22; *Harlow's Weekly*, January 22, 1916, 4; Bizzell, *Farm Tenantry*, 161, 185; Scales and Goble, *Oklahoma Politics*, 65-66, 73-74.

5. Bizzell, *Farm Tenantry*, 118-19, 154, 159; Spillman and Goldenweiser, "Farm Tenantry," 323; testimony of Patrick S. Nagle, *Industrial Relations: Final Report and Testimony Submitted to Congress by the Commission on Industrial Relations Created by the Act of August 23, 1912*, vol. 10 (Washington, D.C.: GPO, 1916), 9076-77 (hereafter *CIR Final Report*).

6. The average acreage of Oklahoma rented and sharecropped land—50.9 and 40.5 acres, respectively—was twice that of the richer soil Mississippi farms, at 22.7 and 18.4 acres, and still produced a lower yield. See Rosen, "Peasant Socialism," 153; Bizzell, *Farm Tenantry,* 156-57, 232-33, 260; Charles S. Johnson, Edwin R. Embree, and William Alexander, *The Collapse of Cotton Tenancy* (Chapel Hill: Univ. of North Carolina Press, 1935), 16-17, 22-23; Charles Holman, "The Tenant Farmer: Country Brother of the Casual Worker," *Survey* 34 (April 17, 1915): 64; W. W. Pannell, "Tenant Farming in the United States," *International Socialist Review* 16 (January 1916): 431-32; Oran Burk, "From a Cotton Picker," *International Socialist Review* 14 (May 1914): 690; "Emptying the School to Work the Farm," *Survey* 38 (September 29, 1917): 576.

7. J. Green, *Grass-Roots Socialism,* 21; Scales and Goble, *Oklahoma Politics,* 68, 71-73, 84; Womack, "Green Corn Rebellion," 72; *Harlow's Weekly,* October 2, 1915, 267-68.

8. McWilliams, *Ill Fares the Land,* 93; Jamieson, *Labor Unionism,* 263; Hall, "Labor Struggles in the Deep South," 183, 205-206, 218-19; Saposs, *Left-Wing Unionism,* 167; Womack, "Green Corn Rebellion," 139; Donald E. Winters, Jr., *The Soul of the Wobblies: The IWW, Religion, and American Culture in the Progressive Era, 1905-1917* (Westport, Conn.: Greenwood Press, 1985), 22-23, 62-63; Eric Hobsbawm, *Primitive Rebels: Studies in Archaic Forms of Social Movements in the Nineteenth and Twentieth Centuries* (New York: W. W. Norton, 1959), 27; Richard Maxwell Brown, "Historical Patterns of American Violence," in *Violence in America: Historical and Comparative Perspectives,* rev. ed., ed. Hugh Davis Graham and Ted Robert Gurr (Beverly Hills: Sage Publications, 1979), 23.

9. McWilliams, *Ill Fares the Land,* 97; Foner, *IWW History,* 258-59; Dubofsky, *We Shall Be All,* 9, 13.

10. Carpenter, "Farmer and Hired Hand," 4-5; William Mead, "Keep Out the Farmer," *Solidarity,* January 11, 1913.

11. Covington Hall, "The Working Farmer and the IWW," *Solidarity,* March 27, 1915; Covington Hall, "The Rebel Farmers of the South," *Industrial Worker* (Seattle), January 23, 1913; *Industrial Worker* (Seattle), January 2, March 6, 1913.

12. *Industrial Worker,* January 30, February 13, March 20, May 1, 1913; Ernest Griffeath, "Can We Unite Both Master and Slave?" ibid., February 6, 1913; *Solidarity,* February 22, 1913, March 20, July 10, 1915.

13. *Industrial Worker,* March 6, 1913; T. F. G. Dougherty, "Advocates a Farmers' Auxiliary," ibid., February 20, 1913; B. E. Nilsson, "Can Tenant Farmers Be Admitted to Membership?" ibid., February 6, 1913.

14. Hall, "Labor Struggles in the Deep South," 219.

15. Ibid., 219-20; WCU membership card, in Box 26, Folder 2, Cole Collection, WHC/OU.

16. Untitled WCU flyer issued from National Headquarters, Lock Box 72, Van Buren, Arkansas, in Correspondence Binder 1-216, 103, Box 36, Folder 2, Cole Collection, WCU/OU; Hall, "Labor Struggles in the Deep South," 220.

17. A Virginia Elliot of Henryetta is the source of the misinformation about Dr. LeFevre. Elliot claimed to have organized for the WCU, but she also said agents of anarchist Emma Goldman took over the union. Elliot, however, seemed obsessed with Goldman and even declared that she was often mistaken for the well-known anarchist. LeFevre was from Pine Bluff and served as one of two delegates from Arkansas to the 1904 SP national convention. The other was Thomas Hagerty. *The Daily Oklahoman*, September 25, October 7, 1917; *Proceedings of the National Convention of the Socialist Party, Chicago, Ill., May 1 to May 6, 1904* (Chicago: Socialist Party, 1904), 38, 45, 301.

18. Land values rose by 246 percent in the period, making it virtually impossible for tenants to buy land. Spillman and Goldenweiser, "Farm Tenantry," 337; Bizzell, *Farm Tenantry*, 153, 167-68; Scales and Goble, *Oklahoma Politics*, 80-81; *Harlow's Weekly*, October 30, 1915.

19. Jamieson, *Labor Unionism*, 261-62; J. Green, *Grass-Roots Socialism*, 81, 301-303; untitled four-page WCU pamphlet, in Correspondence Binder 1-216, 105, Cole Collection, WHC/OU; *Harlow's Weekly*, February 26, March 11, 1916.

20. Cole was an assistant federal attorney at the time and took part in the prosecution. The "Jones Family" label came from the press and the prosecution. It was never used by the defendants themselves, all of whom were regular WCU members. This error has been perpetuated by James Green and John Thompson among others, as well as Bush, Warrick, and Womack. Only Michael Mullen has noted that the defendants never used the "Jones Family" label. A full discussion of the antidipping campaign is in J. Stanley Clark, "Texas Fever in Oklahoma," *Chronicles of Oklahoma*, 24 (winter 1951/52): 429-43; "Jones Family" trial notes, in Correspondence Binder 1-216, 47, 52, 79, 129, Cole Collection, WHC/OU. Michael Mullen, "No Time to Quibble: The Jones Conspiracy Trial of 1917," *Chronicles of Oklahoma* 59 (spring 1982): 224-36; *Oklahoma Labor Unit*, November 6, 1915; *Daily Oklahoman*, December 15, 1915; *Harlow's Weekly*, September 25, 1915; January 1, 15, March 25, 1916; *Eufala Indian Journal*, January 21, 1916.

21. Night riding was a common form of protest, especially against corporate control. For example, Kentucky tobacco farmers used the tactic against the American Tobacco Company in the early 1900s. A secret society of farm laborers and tenants took part in night riding in New Madrid, Missouri, at the same time as the WCU, but the Missouri riders were Democrats, not Socialists, and apparently gained less notoriety as a result. Graham notes night riders were not antiblack per se but focused both on those African-American sharecroppers who undercut white tenants and on those white tenants who did likewise. Civil War antidraft resistance was particularly strong in the Ozarks, the region from which many Oklahoma tenants originally came. Interestingly, swearing oaths on the Bible and a pistol may have older English roots. British historian E. P. Thompson noted early nineteenth-century working-class movements did precisely the same thing in an oath-taking ritual known as *taisez-vous*, meaning "you be silent." For hill country resistance to the Confederate draft and the resisters' use of oaths, secret signs, and countersigns, see Georgia Lee Tatum, *Disloyalty in the Confederacy* (Chapel Hill: Univ. of North Carolina Press, 1934), 36-39, 58-59, 90-91. On the Moffett strike, which sought to raise wages from $1.00 to $1.25, see Jamieson, *Labor Unionism*, 263; E. P. Thompson, *The Making of the English Working Class* (New York: Vintage Books, 1966), 511; Brown, "Historical Patterns," 38-39; J. Green, *Grass-Roots Socialism*, 335; Graham, "Red, White, and Black," 154; Smith, "Sabotage," 8; "Jones Family" trial notes, Correspondence Binder 1-216, 26-27, 30, 49, 55, 77, 125-26, Cole Collection, WHC/OU; Nagle testimony, *CIR Final Report*, 9076-77; *Lexington Leader*, November 27, 1908; *Industrial Democrat*, October 22, 1910; *Daily Oklahoman*, December 9, 1915.

22. Victor Harlow compared the Wobblies and the WCU, but admitted the WCU, unlike the IWW, was a secret society. *Harlow's Weekly*, December 11, 1915; March 11, 1916.

23. Ibid., August 15, 1917; Joan M. Jensen, *The Price of Vigilance* (New York: Rand McNally, 1968), 69-70.

24. Of the earlier chroniclers, only Garin Burbank expressed more doubt about Munson's IWW connections. "Indictments, 1918-1919," a carbon-copy typescript in Box 135, Folder 4, in IWW Collection, ALUA/WSU, gives a fairly complete list of Wobblies under indictment or jailed in Oklahoma and elsewhere. Munson is not among the names listed. Munson *is* listed in the Chicago indictments of 1918, but so were many other non-Wobblies. Burbank, *When Farmers Voted Red*, 146; Bush, "Green Corn Rebellion," 10, 12, 17; Womack, "Green Corn Rebellion,"

1, 87; Patrick Renshaw, *The Wobblies: The Story of Syndicalism in the United States* (Garden City, N. Y.: Doubleday, 1968), 156; J. Green, *Grass-Roots Socialism*, 357–58, 392; *Daily Oklahoman*, October 31, 1917.

25. *Industrial Solidarity*, March 25, 1922, quotes the church council's report. Redmond S. Cole to F. H. Duehay, president of U.S. Board of Parole, December 18, 1918, in Box 30, Folder 1, Cole Collection, WHC/OU; *Daily Oklahoman*, November 1, 1917.

26. "Testimony of William Hoobler," in Transcript of Record, United States Court of Appeals, Eighth Circuit, No. 5170, *Clure Isenhour et al., plaintiff in error, v. United States, defendant in error*, June 1, 1918, 362, 413. Copy in Manuscript and Archives Division, Oklahoma Historical Society, Oklahoma City (hereafter Transcript, *Isenhour et al.*); "Cross-Examination of Tobe Simmons," "Jones Family" trial notes, Correspondence Binder 1–216, 94, Cole Collection, WHC/OU.

27. *Solidarity*, June 9, 1917.

28. The attacks increased in the wake of the Bisbee, Arizona, deportations of striking copper miners. Anton Johannson of the National Labor Defense Council told the OSFL's 1917 convention that the labels IWW or WCU were used regularly by employers to attack any strikes. See *Fourteenth Convention Proceedings of the Oklahoma State Federation of Labor, September 17–20, 1917, Oklahoma City*, OSFL Collection, WHC/OU, Boxes 70 and 88, 71, 73; *Daily Oklahoman*, June 2, 6, 13, 1917; *McAlester News Capital*, August 26, 1917; U.S. Att. John A. Fain to U.S. Att. Gen. Thomas W. Gregory, October 18, 1918, twelve-page typescript, in Box 30, Folder 1, Cole Collection, WHC/OU (hereafter Fain to Gregory), 4.

29. The Chicago charges against Reeder were dismissed on June 20, 1918, according to George, *IWW Trial*, 58–59; Fain to Gregory, 1–3; "Jail and Penitentiary Calendar," unnumbered typescript, Box 135, Folders 4 and 7; White, "Wichita Indictments," 88–91.

30. Fain to Gregory, 1–3; *Annual Report of the Attorney General of the United States for Fiscal Year 1923* (Washington, D.C.: GPO, 1923), 373.

31. The prosecution relied heavily on testimony from the informer, Otis Radcliffe, who claimed to have been a former Wobbly. Fain to Gregory, 1, 4, 12; *Annual Report of the Attorney General of the United States for Fiscal Year 1918* (Washington, D.C.: GPO, 1918), 53; *Annual Report of the Attorney General of the United States for Fiscal Year 1922* (Washington, D.C.: GPO, 1922), 411, 413, 453; *New Solidarity*, December 7, 1918; White, "Wichita Indictments," 298–99, 311.

32. Munson seems to have been released briefly in July, but he was rearrested before the rebellion broke out. John E. Wiggins was finally

sentenced to four years in Leavenworth for draft obstruction in April 1919. He had been held in the federal jail in Muskogee for nineteen months, apparently without charges. Womack, "Green Corn Rebellion," 98; *Daily Oklahoman*, May 30, June 4, 5, 1917; *Seminole County News*, May 31, 1917; *Enid Daily Eagle*, September 24, 1917; *Kansas City Times*, September 25, 1917; *New Solidarity*, April 26, 1919; reports of DOJ agent T. F. Weiss, Oklahoma City/McAlester, September 18, 24, 27, 1917, *USMIR Microfilm*, Reel 6, ff. 279-83; Shirk, *Oklahoma Place-Names*, 135-36.

33. The *Holdenville Times Democrat* of August 10, 1917, declared without substantiation that the IWW led the rebellion. For the aftermath, see Irwin Tucker, "Oklahoma Farmers in War on Bankers and Landlords," *American Socialist*, August 25, September 1, 1917; *Holdenville Times Democrat*, August 10, 1917; *McAlester News Capital*, August 3, 18, 1917; *Daily Oklahoman*, August 4, 22, September 26, October 23, 25-28, 31, November 1-2, 1917; *Ada Weekly News*, August 9, 1917; Wyatt Smith to Rep. Tom McKeown, Fourth District, September 25, 1917, *DOJ/IWW Microfilm*, Reel 5, ff. 81-82; McKeown to U.S. Atty. W. McGinnis, September 29, 1917, ibid., f. 79; McGinnis to William L. Frierson, October 4, 9, 17, 1917, ibid., ff. 74-75, 68-72, 65-66; Frierson to McGinnis, October 4, 11, 17, 1917, ibid., ff. 76-77, 73, 67; Asst. U.S. Attorney General Marvin Underwood to McGinnis, August 29, 1917, ibid., ff. 83-84; unsigned postcard to McGinnis, August 6, 1917, ibid., f. 99.

34. *Enid Morning News*, September 25, 1917; *Daily Oklahoman*, August 4, 5, September 23, 1917; Transcript, *Isenhour et al.*, 390-92, 945, 1007, 1021; White, "Wichita Indictments," 298-99, 311.

35. White, "Wichita Indictments," 298-99, 311.

36. Saposs, *Left-Wing Unionism*, 168-69; Richard Maxwell Brown, "The History of Extralegal Violence in Support of Community Values," in *Violence in America: A Historical and Contemporary Reader*, ed. Thomas Rose (New York: Vintage Books, 1970), 90-91.

Chapter 5. War and Repression

1. Dubofsky, *We Shall Be All*, 349-50.

2. Kirby at the time was head of the anti-union National Association of Manufacturers. His successor, bicycle magnate George Pope, continued Kirby's views by calling unions "lawless organizations" that had to be eliminated. Kirby is quoted in Robert H. Wiebe, *Businessmen and Reform: A Study of the Progressive Movement* (Cambridge: Harvard Univ. Press, 1962), 169; David M. Kennedy, *Over Here: The First World War and American Society* (New York: Oxford Univ. Press, 1980), 70-71, 78-79.

3. J. Green, *World of the Worker*, 87.

4. For a full discussion of war mobilization in Oklahoma, see O. A. Hilton, "The Oklahoma Council of Defense and the First World War," *Chronicles of Oklahoma* 20 (March 1942): 18-42. For the shift from radicalism to conservatism in agriculture, see Kennedy, *Over Here*, 121-23; on the government support of the Farm Bureau, J. Green, *Grass-Roots Socialism*, 346, 351; for the impact of the Federal Farm Loan Act of 1916, Scales and Goble, *Oklahoma Politics*, 52, 68.

5. Burl Noggle, *Into the Twenties: The United States from Armistice to Normalcy* (Urbana: Univ. of Illinois Press, 1974), 93-94; White, "Wichita Indictments," 4-5.

6. Despite the high rate of IWW registration, some federal officials feared the union was using members to organize the army itself. See T. S. Allen, U.S. Attorney for Nebraska, to Atty. Gen. Thomas Gregory, September 17, 1918, *DOJ/IWW Microfilm*, Reel 5, f. 0949; Dubofsky, *We Shall Be All*, 354-55; Preston, *Aliens and Dissenters*, 90.

7. M. A. Hathaway to Arthur Boose, April 8, 11, 1917, *USMIR Microfilm*, Reel 6, 295; Arthur Boose letter in Report of John A. Walen, DOJ agent, September 8, 1917, ibid., 292.

8. Roy Talbert suggests the authoritarian and "morally pure" aspects of Progressivism led supporters to place the needs of the state over those of individuals. Roy Talbert, Jr., *Negative Intelligence: The Army and the American Left, 1917-1941* (Jackson: Univ. Press of Mississippi, 1991), 68; Joan M. Jensen, *Army Surveillance in America, 1775-1980* (New Haven: Yale Univ. Press, 1991), 138-42; Sanford M. Jacoby, *Employing Bureaucracy: Managers, Unions, and the Transformation of Work in American Industry, 1900-1945* (New York; Columbia Univ. Press, 1985), 136-37; Peter Karsten, "Armed Progressives: The Military Reorganizes for the American Century," in *The Military in America: From the Colonial to the Present*, ed. Peter Karsten (New York: Free Press, 1980), 229-71; Preston, *Aliens and Dissenters*, 54-56.

9. Preston, *Aliens and Dissenters*, 60; Dubofsky, *We Shall Be All*, 376-77; F. Thompson and Murfin, *First Seventy Years*, 123-24.

10. Hilton, "Oklahoma Council," 19, 35, 38; Jensen, *Price of Vigilance*, 66-67; Dubofsky, *We Shall Be All*, 399-402, 404-405.

11. The Tulsa APL's directors were John A. Hammer, C.E. Lahman, and L. Bartlett. It investigated eleven cases involving the Espionage Act and twenty-one enemy alien cases. Lampe, *Tulsa County*, 58-59, 61, 152-54; Hilton, "Oklahoma Council," 18, 38; Jensen, *Price of Vigilance*, 125.

12. Lampe, *Tulsa County*, 152; Reports of T. Weiss, DOJ agent for Oklahoma City, November 8, 9, 1917, *USMIR Microfilm*, Reel 6, ff. 267, 270; *Daily Oklahoman*, November 1, 1917; Jensen, *Price of Vigilance*, 59.

13. Jensen, *Price of Vigilance*, 59; Dubofsky, *We Shall Be All*, 359-60; Applen, "Migratory Harvest Labor," 160.

14. *Oklahoma Farmer-Stockman*, May 25, 1917; *Daily Oklahoman*, May 31, June 3-4, 10, 1917; "Harvest Labor," *New York Times*, July 18, 1917.

15. *Solidarity*, June 9, 16, July 7, 1917.

16. *Drumright Evening Derrick*, May 28, 29, 31, June 4, 6, 1917; *Enid Daily Eagle*, June 13, 15, 1917; White, "Wichita Indictments," 244.

17. The *Drumright Evening Derrick* claimed the WCU and IWW had "amalgamated," which was clearly untrue, as chap. 4 has demonstrated. *Drumright Evening Derrick*, June 1, 4, 1917; D. Earl Newsom, *Drumright! The Glory Days of a Boom Town* (Perkins, Okla.: Evans Publications, 1985), 89-90.

18. The Bisbee and Jerome deportations are well documented and discussed in several works. The Kansas City hall was the target of several attacks, especially an April 3 raid by off-duty Marines and state militia that presaged later raids nationwide. Dubofsky, *We Shall Be All*, 383-86, 405-408; A. S. Embree, "Introduction," in Harrison George, *The IWW Trial* (Chicago: Industrial Workers of the World, 1919), 3-13; Alexander M. Bing, *War-Time Strikes and Their Adjustment* (New York: E. Dutton, 1921), 255-58, 263; *New York Times*, July 13, 1917; *Industrial Worker* (Seattle), June 23, 1917; *Solidarity*, April 7, June 16, 23, July 14, 1917.

19. F. Thompson and Murfin, *First Seventy Years*, 144-45; letter from Phineas Eastman, Augusta, Kansas, May 21, 1917, quoted in *Evidence and Cross-Examination of William D. Haywood*, 223.

20. *Solidarity*, June 16, 1917; U.S. Attorney General Harry Dougherty to President Warren G. Harding, "In the Matter of Execution of Executive Clemency," undated memorandum, in *DOJ/IWW Microfilm*, Reel 14, ff. 903-905; White, "Wichita Indictments," 57, 61. White's source is Gallagher's prison file at Leavenworth. See also *United States Attorney General's Parole Report*, February 9, 1920.

21. "Oil Workers Wanted in Oklahoma," *Solidarity*, July 7, 1917; *Industrial Worker*, July 7, 1917; Fred Edgecombe to J. Gallagher, August 4, 1917, typed copy in *USMIR Microfilm*, Reel 6, 298; J. Gallagher to editor, *Solidarity*, August 11, 1917; James Morton Smith, "Criminal Syndicalism in Oklahoma: A History of the Law and Its Application" (master's thesis, University of Oklahoma, 1946), 29.

22. *Tulsa Democrat*, August 10, 1917; *Drumright Daily News*, November 2, 1917; *Daily Oklahoman*, November 2, 1917; *Tulsa World*, August 3, 1917; report of federal agent Charles G. Findley, Oklahoma City, August 31, 1917, in *USMIR Microfilm*, Reel 6, f. 301.

23. The men expelled were W. H. Walton, later an IWW job delegate and victim of the Tulsa Outrage, and Charles Walsh. Findley report, *USMIR Microfilm*, Reel 6, ff. 302-303; report of federal agent John A. Whalen to Military Intelligence Division, September 8, 1917, ibid., f. 286; Phineas Eastman to J. Ryan, August 13, 1917, typed copy in ibid., f. 294; *Tulsa World*, August 2, 1917; *Solidarity*, August 18, 1917; *Industrial Worker*, August 18, 1917; "Bills Drafted to Curb the IWW," *Survey* 38 (August 25, 1917): 457-58.

24. James H. Fowler, Jr., "Creating an Atmosphere of Suppression, 1914-1917," *Chronicles of Oklahoma* 59 (summer 1981): 202-23; *Daily Oklahoman*, April 25, 1914; Virginia C. Pope, "The Green Corn Rebellion: Case Study in Newspaper Self-Censorship," (master's thesis, Oklahoma A&M College, 1940), 37; Hilton, "Oklahoma Council," 20.

25. Congdon also served as state chair of the "Four-Minute Men" and, only days after the Tulsa Outrage, went to Europe on a special war mission. *Tulsa World*, August 6, 7, 9, November 10, 1917; Lampe, *Tulsa County*, 139, 202, 234.

26. The stickerettes were printed in two colors on gummed paper. Barely larger than postage stamps, they came in packages of a hundred or so and cost ten cents. Most found their way onto boxcars, lampposts, tools, and bunkhouses. "At the peak of the stickerette campaign it was said that every Boxcar in the country carried with it at least one good argument in favor of joining the I.W.W.," Ralph Chaplin noted. See Chaplin, *Wobbly*, 194; Bernard A. Weisberger, "Here Come the Wobblies!" *American Heritage* 18 (June 1967): 89; *Harlow's Weekly*, June 6, 1917.

27. William D. Stephens to Woodrow Wilson, July 9, 1917, and Thomas W. Gregory to Wilson, August 21, 1917, cited in David M. Kennedy, *Over Here: The First World War and American Society* (New York: Oxford Univ. Press, 1980), 88; F. Thompson and Murfin, *First Seventy Years*, 124; Dubofsky, *We Shall Be All*, 405-408; reports of federal agent John Whalen to Military Intelligence, September 6, 8, 1917, *USMIR Microfilm*, Reel 6, ff. 287-89; *Industrial Worker*, September 12, 1917; *New York Times*, September 29, 1917.

28. Information on Harry Trotter's Oklahoma activities is in *Solidarity*, April 14, May 19, 26, 1917. Trotter apparently worked with W. A. Gourland, who was recruiting Oklahoma City workers for Railroad

Workers Industrial Union no. 600. George, *IWW Trial*, 10-12; *Defense News Bulletin* 30 (June 8, 1918), 4, in Box 181, Folder 16, IWW Collection, ALUA/WSU; White, "IWW and the Mid-Continent," 69; report of agent Whalen, September 30, 1917, *USMIR Microfilm*, Reel 6, f. 277; *Solidarity*, October 13, 1917.

29. At least one fellow worker tried to warn Boose of his arrest. See letter from Charles Schultz to Willam Haywood, September 22, 1917, quoted in *Evidence and Cross-Examination of William D. Haywood*, 177. The Boose quotation is from Stuart H. Holbrook, "The Last of the Wobblies," in *Little Annie Oakley and Other Rugged People* (New York: Macmillan, 1948), 172-73; *Daily Oklahoman*, October 23, 1917.

30. Anderson, on October 27, 1917, wrote E. F. Doree that organizing the oil fields was important because "everybody uses oil." Quoted in Attorney General Dougherty to President Harding, "In the Matter of the Application for Pardon of C. W. Anderson," undated fifteen-page letter in Pardon Attorney Files, Federal Record Group 204, in *DOJ/IWW Microfilm*, Reel 14, ff. 527, 529, 531; *Solidarity*, October 27, 1917; White, "Wichita Indictments", 59-60, 64, 67-68, 77; F. Thompson and Murfin, *First Seventy Years*, 125.

31. For a discussion of the Gulf Goast strikes, see Allen, *Texas Labor*, 222-27; *Solidarity*, October 6, 1917.

32. *Industrial Worker* (Seattle), November 16, 1911; "In re Francik," undated letter from Attorney General Dougherty to President Harding, *DOJ/IWW Microfilm*, Reel 14, ff. 582-84; unsigned memo to Dougherty, October 14, 1922, ibid., ff. 859-60; Dougherty to Pardon Attorney James Finch, December 5, 1922, ibid., f. 883; White, "Wichita Indictments," 60-61, 264.

33. Gallagher (as John Shannon) to Gordon, October 24, 1917, quoted in statement of Caroline Lowe, attorney, Wichita case, *DOJ/IWW Microfilm*, Reel 214, f. 889; Gallagher to Anderson, November 8, 1917, copy in letter to Finch, ibid., 914; Gallagher to Anderson, October 31, 1917, from third Wichita indicitment, quoted in White, "Wichita Indictments," 65, 71.

34. *Solidarity*, October 27, 1917; *Daily Oklahoman*, October 21, 22, 1917; *Tulsa World*, October 30-31, 1917; telegram from Meserve to Attorney General Gregory, October 29, 1917, *DOJ/IWW Microfilm*, Reel 5, f. 64; Lampe, *Tulsa County*, 73.

35. *Drumright Daily News*, October 30, 1917, identified Powers as an OWIU organizer in the Cushing Field. It also claimed the IWW was tied to the Texas strike, but federal agent C. Procter told Attorney General

Dougherty there was no connection to the IWW. Procter, however, said he believed the German spies were now infiltrating the AFL because the IWW's "notorious reputation" made it difficult to use as a cover. Procter to Gregory, October 29, 1917, in Dubofsky, *Strikes Files*, Reel 9, 216-25; Lampe, *Tulsa County*, 73; *Tulsa Times*, October 30, 1917; *Tulsa World*, October 31, November 6, 1917.

36. Federal agent John Whalen doubted the IWW connection to the Pew case, although he had no doubts the IWW was planning "mischief" in the oil fields. Weiss confirms city authorities acted on their own despite the requests of Whalen and Deputy U.S. Marshal John Moran. Weiss said "the time was not ripe" for arrests. Reports of agent Whalen, October 30, November 3, 1917, *USMIR Microfilm*, Reel 6, f. 270; report of agent Weiss, November 8, 1917, ibid., ff. 274-75; quoted in NCLB, "'Knights of Liberty' Mob."

37. The men arrested were identified as E. M. Boyd, age 53; John Doyle (or Boyle), 48; John Fitzsimmons, 34; Joe French, 33; J. R. Hill, 34; Gunnard Johnson, 26 (misidentified both as "Gerard" and as "Bernard); Tom McCaffrey, 33; Bob McDonald, 27; John McCurry, 29; John Myers, 34; and Charles Walsh, 27. Except for French, who was French, and Johnson, a Swede, the men were either American or British. French said the vagrancy charges were ridiculous as the men all wore good clothes and carried money. Weiss said the men appeared to be engaged only in normal job organizing. Information seized in the raid revealed only that the IWW had organizers in Healdton, Cushing, and Butler County, Kansas, none of which was startling news. See *Daily Oklahoman*, November 9, 1917; *Tulsa World*, November 6, 1917; Weiss report, November 8, 1917, *USMIR Microfilm*, Reel 6, f. 271; November 9, 1917, 267; Whalen report, November 17, 1917, ibid., 264-65; "Testimony of Joe French," in George, *IWW Trial*, 93-94.

38. Several sources seem to believe that twelve men were arrested and tried, but I have found no confirmation that Morris was arrested, let alone tried with the others. Weiss report, November 8, 1917, ibid.; Whalen report, November 17, 1917, ibid.; *Tulsa World*, November 6, 7, 1917.

39. *Tulsa World*, November 7, 9, 1917.

40. Ibid., November 9, 1917.

41. Most sources have claimed seventeen men were whipped. The correct total is sixteen, which can be ascertained by counting the names of the victims mentioned in the NCLB report. In addition to the ten previously named the others were W. H. Walton, L. R. Mitchell, Thomas Fischer, Gordon Dimison, J. Ryan, and Jack Sneed, who was not a

Wobbly. Lampe claims the men faced jail sentences as well, which Evans offered to suspend. But no other sources indicate this, and in all likelihood Lampe is simply wrong. Lampe, *Tulsa County*, 221-22; "Mob Violence in the United States," *Survey* 40 (April 27, 1918): 102; "Testimony of Joe French," in George, *IWW Trial*, 93-94; NCLB, "'Knights of Liberty' Mob," 14-15; *Tulsa World*, November 10, 1917; *Tulsa Democrat*, November 10, 1917; *Tulsa Times*, November 10, 1917; "Vigilantes Mob IWW in Oklahoma," *Industrial Worker*, November 17, 1917.

42. Tarring and feathering seems to have been a common way of mistreating radicals in Oklahoma in this period. There is some dispute, however, as to what was used to whip the Tulsa Wobblies. The article in *Survey* on April 27, 1918, as well as contemporary accounts such as the NCLB's, said either rope or new rope, soaked in salt water. Chaplin said it was blacksnake whips, while French, who was a victim, said it was half-inch pipe. Interestingly enough, the name "Knights of Liberty" originally belonged to a dissident Copperhead Democrat group opposed to Lincoln during the Civil War. The group was heavily infiltrated by Pinkerton agents. See Hilton, "Oklahoma Council," 36, and James H. Fowler, Jr., "Tar and Feather Patriotism: The Suppression of Dissent in Oklahoma During World War I," *Chronicles of Oklahoma* 56 (winter 1978-79): 421-22. Jensen, *Army Surveillance*, 25; NCLB, "'Knights of Liberty' Mob," 6-7; "Testimony of Joe French," in George, *IWW Trial*, 94; *Tulsa World*, November 10, 1917; "Who Is to Blame for Tulsa?" *Industrial Worker*, November 24, 1917; Hilton, "Oklahoma Council," 36; Lampe, *Tulsa County*, 73; Chaplin, *Wobbly*, 211.

43. *Industrial Worker*, February 9, 1918, said the mob included J. Edgar Pew; City Attorney John Meserve, who was accused of whipping most of the men; local political leader Buck Lewis; Tulsa Police Chief E. L. Lucas; and five police officers, including H. H. Townsend. Those five were accused of killing a young man named Vernon Hatfield in the spring of 1917. The article also charged that Glenn Congdon, editor of the *Tulsa World*, was not simply an observer but actually a mob member. White notes that NCLB investigator L. A. Brown, in a letter to NCLB official Roger Baldwin, said two policemen named Patten and Carmicle were in the mob and that Judge Evans and Meserve at least knew of the mob beforehand. *Industrial Solidarity*, June 10, 1922, identified LaFallete as a mob member. NCLB, "'Knights of Liberty' Mob," 8-10, 13-14; *Tulsa Democrat*, November 11, 1917.

44. NCLB, "'Knights of Liberty' Mob," 14-15; *Tulsa World*, November 10-11, 1917; *Tulsa Democrat*, November 10-11, 1917; "Mob Violence in

America," 102; "Vigilantes Mob IWW in Oklahoma," *Industrial Worker*, November 17, 1917; *Daily Oklahoman*, November 11, 1917; *Oil and Gas Journal*, November 15, 1917.

45. Assistant U.S. Attorney Cole mentioned Macklin's arrest, but he admitted Macklin did not engage in antiwar activities. Interestingly enough, Cole also informed the attorney general that federal grand juries failed to indict any of the men. Redmond S. Cole to Attorney General Gregory, January 24, 1918, in Correspondence Binder 1-224, Box 36, Folder 3, Cole Collection, WHC/OU; *Drumright Daily News*, November 4, 12, 1917; *Drumright Evening Derrick*, November 7, 1917; report of agent Whalen, November 18, 1917, *USMIR Microfilm*, Reel 6, f. 254; report of agent Weiss, November 9, 1917, ibid., 267-69; *Tulsa World*, November 11, 1917; *Daily Oklahoman*, November 9, 12, 1917; White, "Wichita Indictments," 66-67.

46. *Tulsa World*, November 12, 1917; *Daily Oklahoman*, November 12, 1917; *Drumright Evening Derrick*, November 7, 1917; *Oil and Gas Journal*, November 8, 15, 1917.

47. "What Do You Think of This?" (Chicago: Industrial Workers of the World, 1918), copy in "IWW-Persecution, Tulsa," in vertical files, Labadie Collection, University of Michigan, Ann Arbor; *Industrial Worker*, December 1, 1917; *Oklahoma Federationist*, November 10, 17, 24, 1917; *Drumright Daily News*, November 23, 1917.

48. The arrest may have been as early as November 2. *Drumright Daily News*, November 3, 15, 26, 1917; *Tulsa World*, November 10, 1917; *Industrial Worker*, November 24, 1917; *Daily Oklahoman*, November 16, 17, 27, 1917; *Drumright Evening Derrick*, November 18, 26, 1917; John Womack, Sr., *Annals of Cleveland County, Oklahoma, 1889-1957* (Norman: Transcript Press, 1976), 41; Atty. Gen. Thomas Gregory to W. McGinnis, November 28, 1917, *DOJ/IWW Microfilm*, Reel 5, ff. 54-56; McGinnis to Gregory, November 28, 1917, ibid., ff. 54-56.

49. Sapper was originally arrested November 12 solely for "lodge activities." He was later charged with espionage, charges that Assistant U.S. Attorney General O'Brien questioned. *Daily Oklahoman*, November 9, 17, 19, 21-23, 1917; *Industrial Worker*, November 24, December 22, 1917; C. W. Anderson, "Oil Field Raids," *Defense News Bulletin* 9 (January, 1918); *New York Times*, November 21, 1917; *Drumright Evening Derrick*, November 17, 1917; "Memorandum of the Attorney General to the Acting Pardon Attorney," October 14, 1920, *DOJ/IWW Microfilm*, Reel 14, ff. 579-80; W. McGinnis to Attorney General Gregory, December 5, 1917, *DOJ/IWW Microfilm*, Reel 5, ff. 46-53. *DOJ/IWW Microfilm* Reel 14 contains files from

National Archives Record Group 204, Casefile 39-242 of the U.S. Pardon Attorney; White, "Wichita Indictments," 83-84, 101-103; 264; Redmond S. Cole to Attorney General, January 24, 29, March 9, 1918, and John Lord O'Brien, Special Assistant to Attorney General, to Cole, February 8, 1918, in Correspondence Binder 1-224, Box 36, Folder 3, Cole Collection, WHC/OU; American Civil Liberties Union, "The Truth about the IWW Prisoners" (New York: American Civil Liberties Union, 1920), 21-22, copy in vertical files, ALUA/WSU.

50. According to Cole's notes and letters, Macklin's case was Federal Criminal Case No. 1235 for the western district of Oklahoma, while Edgecombe's was Federal Criminal Case No. 1288 for the same district. Sapper's Oklahoma espionage case (Federal Criminal Case No. 1247) was also dropped. See Cole to Attorney General, March 9, 1918, Correspondence Binder 1-224, Box 36, Folder 3, Cole Collection, WHC/OU. *Otter Valley Socialist*, February 28, 1918.

51. William E. Forbath, *Law and the Shaping of the American Labor Movement* (Cambridge: Harvard Univ. Press, 1991), esp. chap. 4; Victoria C. Hattam, *Labor Visions and State Power: The Origins of Business Unionism in the United States* (Princeton, N.J.: Princeton Univ. Press, 1993), ix, 3.

52. Philip Taft notes that news reports said 166 persons were indicted, but 1 person had been named twice. Taft, "The Federal Trials of the IWW," *Labor History* 3 (winter 1962): 61-76; George, *IWW Trial*, 58-59; "Testimony of Harry Trotter, August 6, 1918," in transcript of *William D. Haywood et al. v. the United States of America* (hereafter *Haywood v. United States*), 10541-47, typescript copy in Box 116, Folder 1, IWW Collection, ALUA/WSU; American Civil Liberties Union, "Truth about IWW Prisoners," 14, 46; Dubofsky, *We Shall Be All*, 427-33; *New Solidarity*, December 21, 1918.

53. American Civil Liberties Union, "Truth about IWW Prisoners," 16; George, *IWW Trial*, 25; "Testimony of Stanley J. Clark, August 8, 1918," *Haywood v. United States*, 10873-77, 10893, Box 116, Folder 3, IWW Collection, ALUA/WSU; "Testimony of Dan Krich, August 13, 1918," *Haywood v. United States*, 10029, Box 115, Folder 4, IWW Collection, ALUA/WSU; *New Century*, January 6, 1911.

54. Preston, *Aliens and Dissenters*, 130-31; W. P. McGinnis to Thomas Gregory, December 5, 1917, in *DOJ/IWW Microfilm*, Reel 5, ff. 50-53. For McGinnis's expected eagerness, see McGinnis to Gregory, August 20, 1917, *DOJ/IWW Microfilm*, Reel 5, ff. 86-94; telegram from Tulsa County Council of Defense to Gregory, November 26, 1917, *DOJ/IWW Microfilm*, Reel 5, f. 57.

55. Although a grand jury failed to indict Sapper, he was held in federal jail at Oklahoma City based on the Kansas indictment. In contrast to McGinnis in the eastern district, Cole said the U.S. Attorney in the western district would cooperate with Kansas officials. White, "Wichita Indictments," 80-82; Clayton R. Koppes, "The Kansas Trial of the IWW, 1917-1919," *Labor History* 16 (summer 1975): 341-42; American Civil Liberties Union, "Truth about IWW Prisoners," 22; Cole to attorney general, March 9, 1918, Correspondence Binder 1-224, Box 36, Folder 3, Cole Collection, WHC/OU.

56. Winthrop Lane's article is the best description of the jail conditions the prisoners endured. Winthrop D. Lane, "Uncle Sam, Jailer," *Survey* 42 (September 6, 1919): 800-12, 834. Koppes, "Kansas Trial," 343; *New Solidarity*, December 28, 1918; Wencil Francik to editor, *New Solidarity*, May 17, 1919; "Two Years Without Trial," *Detroit News*, September 14, 1919.

57. Koppes notes contemporary news reports suggest the Wobblies feared divulging more "evidence," but he argues they may have simply been showing disdain for the judicial system, as did IWW members on trial in Sacramento. White argues the decision was more likely based on the IWW's lack of funds, an argument I share. Koppes, "Kansas Trial," 353-54; White, "Wichita Indictments," 221-26, 229, 240; *Kansas City Star*, December 14, 1919; *New Solidarity*, December 20, 1919; Caroline A. Lowe, "A Letter from Our Attorney on the Wichita Case," *One Big Union Monthly* 1 (September 1919): 9.

58. White, "IWW and the Mid-Continent," 79; White, "Wichita Indictments," 254-59, 262; *Kansas City Star*, December 19, 1919.

59. *Industrial Solidarity*, February 18, 1922; *New Solidarity*, February 28, 1920; *Industrial Pioneer* 1 (June 1921): 59; *New York Times*, May 28, 1923; Hall, "Labor Struggles in the Deep South," 222.

60. One of those accepting the conditional amnesty was Arthur Boose, whose family was then living in Tulsa. For a full discussion of Gallagher's case, see his Pardon Attorney's file, *DOJ/IWW Microfilm*, Reel 14, f. 902-51; see *Industrial Solidarity*, May 20, 1922, December 29, 1923; *Annual Report of the Attorney General of the United States for Fiscal Year 1923*, 418; Preston, *Aliens and Dissenters*, 262-64; White, "Wichita Indictments," 68; *New York Times*, December 16, 1923.

61. Scales and Goble, *Oklahoma Politics*, 52, 68.

62. Ibid., 64-65; Grant McConnell, *The Decline of Agrarian Democracy* (Berkeley: Univ. of California Press, 1959), 19-20; Rosen, "Peasant Socialism," 215-17.

NOTES TO PAGES 119-21   237

Chapter 6. Trials and Tribulations

1. Charles Alexander in his study of the Klan in the Southwest notes radicals were more often a target in Oklahoma because African Americans represented just 7 percent of the population and foreign-born whites just 2 percent. Whalen reported an alleged December 4, 1917, IWW threat to blow up an Acme Oil refinery at Jennings unless a worker was reinstated. The source of the information was the refinery superintendent. Charles Comer Alexander, "Invisible Empire in the Southwest: The Ku Klux Klan in Texas, Louisiana, Oklahoma, and Arkansas in 1920-1930" (Ph.D. dissertation, University of Texas, 1962), 16, 18-19, 20-21; report of John A. Whalen to the War Department, December 23, 1917, *USMIR Microfilm*, Reel 6, 242-43; *Oil and Gas Journal*, November 15, 1917; "Mob Violence in the United States," 101; *Industrial Worker*, April 27, 1918; *Daily Oklahoman*, May 16, 18, June 6-7, 12, 1918; *New York Times*, September 1, 1918; *Tulsa World*, May 8, 11, 29, 1918; *Oklahoma Federationist*, December 8, 1917; *Drumright Evening Derrick*, December 3, 1917.

2. *Harlow's Weekly*, January 23, 1918, 12; *Tulsa World*, November 1, 1918.

3. Examples of such reports regarding radical activity in Oklahoma can be found in U. G. Worrilow's letters to the director of military intelligence. Worrilow, a retired colonel, served as army recruiting officer for Oklahoma City. His estimates of IWW strength range from 5,000 in April 1920 to only 250 six months later in November, which raises the question of how much effort he actually put into the reports. See Worrilow to director, Military Intelligence, April 21, November 30, 1920, in *USMIR Microfilm*, Reel 9, ff. 296-97, 300-302; J. F. Darby to Lt. Col. Ralph H. Van Deman, head of Military Intelligence, March 21, 1918, *USMIR Microfilm* Reel 6, ff. 244-45; W. McGinnis to Attorney General Gregory, December 5, 1917, *DOJ/IWW Microfilm*, Reel 5, ff. 46-53; Asst. Atty. Gen. William Fitts to McGinnis, February 7, 1918, ibid., ff. 44-43; U.S. Marshall B. A. Enloe Jr., to Attorney General Gregory, January 29, 1918, ibid., 44-45; McGinnis to Gregory, March 27, 1918, ibid., ff. 39-41; Talbert, *Negative Intelligence*, 138, 201.

4. *Daily Oklahoman*, June 8, 1918; *Harlow's Weekly*, December 14, 1917.

5. E. P. Thompson, *Whigs and Hunters: The Origins of the Black Act* (New York: Pantheon, 1975), 259.

6. For examples of Wobbly courtroom behavior, see Dubofsky, *We Shall Be All*, 423-24. But, as Dubofsky notes, some Wobblies, such as Elizabeth Gurley Flynn and Arturo Giovannitti, understood the law better than most. Their use of pretrial challenges in the Chicago case

caused the dismissal of charges against them, which allowed them to avoid the others' fate. But their actions also led to their alienation from and eventual break with the union. Such questions of law are dealt with by E. P. Thompson, who argued that the law binds the rulers to act only in the ways permitted; to do otherwise is often in the hands of the ruled, especially jurors. However, as happened in America in World War I, Thompson observes that a ruling elite may often dispense with the law and its own rhetoric because to obey its own rules would be to surrender its hegemony. E. P. Thompson, *Whigs and Hunters*, 265, 267.

7. "The General Defense Committee of the IWW," *International Socialist Review* 18 (February 1918): 408-409.

8. Federal agent James Findlay said Casey was indeed a Wobbly, based on letters indicating he had contributed two dollars to the union's defense fund. In a related vein, intelligence head Lt. Col.Ralph H. Van Deman apparently believed a felon's allegation that Casey had received the equivalent of one thousand dollars in German money. But Bureau of Investigation chief Bruce Bielaski and agent Findlay found the story improbable. Report of James G. Findlay, agent in charge of Oklahoma City, re Harry Casey, December 18, 1917, and Findlay to Will C. Austin, Department of Justice, Fort Worth, undated, *USMIR Microfilm*, Reel 7, f. 319-25; Van Deman to A. Bruce Bielaski, Chief of Bureau of Investigation, December 18, 1917, and Bielaski to Van Deman, December 28, 1917, *USMIR Microfilm*, Reel 7, ff. 333-34; *Tulsa World*, November 4, 1917; *New York Times*, December 15, 1917; Cole to U.S. Attorney General, January 28, 1918, Cole Collection, WHC/OU.

9. *New Solidarity*, November 30, 1918; "Charles Krieger at Mercy of Standard Oil," *New Solidarity*, May 3, 1919; Eugene Lyons, *Assignment in Utopia* (New York: Harcourt Brace and World, 1938), 14-15; "Framed Up by Standard Oil," *Rebel Worker*, June 15, 1919.

10. "Indictments," typescript in IWW Collection, ALUA/WSU, Box 128, Folder 4; "Standard Oil Frame Up in Tulsa," *New Solidarity*, March 8, 1919; Eugene Lyons, "Tulsa: A Study in Oil," *One Big Union Monthly* 1 (December 1919): 36; *Tulsa World*, May 9, 1918.

11. Lyons, "Tulsa," 36-37.

12. Whalen initially believed that Foster, a former WFM member, was lying about having only just met Vowells (whom Foster knew as "Earl Gordon") and John Hall. Whalen to War Department, December 21, 1917, *USMIR Microfilm*, Reel 6, ff. 229-31; see also Whalen reports for December 22, 1917 (Reel 7, ff. 314-15), where he interviewed J. T. Foster, December 23, 1917, and January 18, 1918 (Reel 7, ff. 326-27, 316-17)

regarding Vowells's wife, Mozer Stricker. Lyons, "Tulsa," 37; *Tulsa World*, November 5, 1917.

13. Lyons, "Tulsa," 37; *Oklahoma Leader*, May 10, 1919; "Standard Oil Frame Up in Tulsa," *New Solidarity*, March 8, 1919.

14. "Standard Oil Frame Up," ibid.; on Vowell's plea, see *Tulsa Times*, August 28, 1918, and *Tulsa World*, August 29, 1918; *New Solidarity*, May 3, 1919.

15. Little information exists about Bonstein, other than that he was Jewish and had offices in the Turner Building in Tulsa. He also seems not to have taken part in the courtroom presentations, though it is probable he did much of the case research. Lyons, *Assignment in Utopia*, 17; *New Solidarity*, May 3, August 30, 1919; *Oklahoma Leader*, May 10, 1919; *General Office Bulletin*, February 14, 1918, copy in Box 31, Folder 1, IWW Collection, ALUA/WSU; White, "Wichita Indictments," 93.

16. Lowe's instructors included Arthur LeSeur, the attorney for the Non-Partisan League who negotiated the 1917 wage agreement with the IWW. *New Solidarity*, May 17, 1919; *Party Builder* 84, October 16, 1912, 4; *Tulsa World*, October 16, 1919; White, "Wichita Indictments," 15n, 113, 196-98.

17. No biography of any sort exists for Moore, but White's biographical sketch, although short, is extremely useful. Lyons, *Assignment in Utopia*, 13; Eugene Lyons, *The Life and Death of Sacco and Vanzetti* (New York: Da Capo, 1970), 66-67; White, "Wichita Indictments," 192-96.

18. Lyons, *Assignment in Utopia*, 17; White, "Wichita Indictments," 22, 113, 269; *Industrial Worker*, December 1, 1917; *Solidarity*, May 22, 1920.

19. Short biographies of Cole are found in Rex F. Harlow, *Makers of Government in Oklahoma* (Oklahoma City: Harlow Publishing Company, 1930), 769; and Lyle H. Boren and Dale Boren, eds., *Who Is Who in Oklahoma* (Guthrie, Okla.: Co-Operative Publishing Company, 1935), 102-103. See also *Tulsa World*, June 7, 1918, on Enfield case; R. S. Cole to Mary Cole, September 27, 1918, and Cole to Prentice Rowe, September 22, 1918, both in Box 37, Folder 2, Cole Collection, WHC/OU.

20. The son of a Kentucky physician and former Kentucky legislator, Moss was a typical southern business progressive. His family were prosperous tobacco farmers and breeders of racehorses. He came to Oklahoma in 1902 and by 1913 was a partner in the Tulsa law firm of Martin and Moss, which represented several corporations. See Joseph B. Thoburn, *A Standard History of Oklahoma*, vol. 3 (Chicago: American Historical Society, 1916), 1123-24; *Tulsa World*, February 26-27, 1919; *New Solidarity*, March 1, April 26, May 3, 1919; *Oklahoma Leader*, May 10, 1919 (quotation).

21. The Lowe quotation is from *Tulsa World*, October 16, 1919; see also *Tulsa World*, March 1, 1919; jury docket, District Court of Tulsa County, June term 1919, *State v. Charles Krieger*, Case no. 144, 5. Copy in Box 37, Folder 4, Cole Collection, WHC/OU; *New Solidarity*, September 20, 1919.

22. A recent account of the incident by D. Earl Newsom uncritically accepts the claim that the IWW instigated the riot and also makes the common, careless error of calling the Wobblies the "International" Workers of the World. H. Malkus quoted city officials as saying twelve hundred Wobblies were in Drumright, clearly an exaggerated number, while Gaylord Wilcox blamed the IWW for the strike and urged the use of state power to stop "that element of unnatural persons." But the arrested workers were all members of the AFL oil workers' union. Although a state arbitration board found for the strikers, Southwestern Bell refused to comply with the decision. The story of the strike and riot is covered in the *Drumright Derrick*, September 21-25, October 2, 13, 1919. For examples of operatives' reports on IWW involvement, see Lt. J. H. Cary, Regimental Intelligence Officer, to Commanding Officer, Second Infantry, Oklahoma National Guard, McAlester, November 8, 9, 1919, and Division Special Officer G. W. Yeartu to J. H. Burnett and D. Van Hecke, November 8, 1919, all in Box 11, Folder 5, Governor James A. B. Robertson Papers, Oklahoma State Archives, Department of Libraries, Oklahoma City (hereafter Robertson Papers, OSA); Chris Madsen to Robertson, November 4, 12, 1919, in Box 11, Folder 9, ibid. For UMW District 14 support of the Wichita IWWs, see "Here Is the Resolution of the United Mine Workers Demanding Fair Play for the I.W.W." flyer signed by District President Thomas Harvey and Secretary Alexander Howat. Copy in "IWW Prosecutions—Kansas" folder in vertical files, Labadie Collection, University of Michigan, and *Oklahoma Leader*, August 16, 1919; Newsom, *Drumright!* 98-100; H. Malkus, "Political Grudge, Bolshevism and History Tear Drumright," *Tulsa World*, September 25, 1919; *Tulsa World*, September 24, 30, October 1, 20, November 5, 1919; O'Connor, *OWIU-CIO History*, 175; Creek County Judge Gaylord R. Wilcox to Robertson, September 27, 1919, copy in Box 3, Folder 22, Robertson Papers, OSA; "Report of State Board of Arbitration and Conciliation," seven-page typescript, and Labor Commissioner Claud Connally to Robertson, October 13, 1919, both in Box 26, Folder 10, Robertson Papers, OSA; Proclamation of Martial Law, December 4, 1919, Box 11, Folder 10, Robertson papers, OSA. The reference to a lawless conspiracy is in "An Open Letter from Governor Robertson to The People of Oklahoma," undated, in Box 26, Folder 7, Robertson Papers, OSA.

23. The *Tulsa World* gave extensive coverage to the case, including interviews and comments not found in the transcript. Despite its anti-radicalism, the paper's reporting of the trial seems relatively balanced. *Tulsa World*, October 8, 9, 14-20, 22-23, 30 1919; *New Solidarity*, October 25, 1919. The full citation is *State of Oklahoma v. Charles Krieger*, Criminal Case No. 1576, County District Court, Tulsa, Oklahoma. A copy of the transcript is in AG Files, OSA.

24. Lyons, *Assignment in Utopia*, 18; Cole to Mary Cole, October 23, 1919, in Box 10, Folder 1, Cole Collection, WHC/OU.

25. Lyons, *Assignment in Utopia*, 15-16.

26. Among the witnesses was *Tulsa World* publisher Eugene Lorton, who surprisingly claimed ignorance about the IWW and its ideology, despite the anti-Wobbly editorials his paper had published in 1917. Another witness, a seventeen-year-old named Currin who taught the prisoners, testified that a social worker named Betty Bettex had told him Carter Oil would give him a car if he said Krieger had confessed. He told the court he believed Krieger was innocent. *Tulsa World*, October 24, 1919; *New Solidarity*, October 18, 1919.

27. *Tulsa World*, October 25, 1919; *New Solidarity*, November 1, 1919.

28. *Tulsa World*, October 26, 29, 1919; *New Solidarity*, November 1, 1919.

29. Cole seems to have worried that further delays would carry the trial past the end of the June term and might force a mistrial, but Oklahoma City Judge N. McNeill assured Cole that Oklahoma law did not require a mistrial. See Cole to McNeill, October 27, 1919, Box 10, Folder 2, Cole Collection, WHC/OU; and McNeill to Cole, October 30, 1919, Box 10, Folder 1, ibid.; *Tulsa World*, October 30, 1919; *New Solidarity*, November 8, 1919.

30. *Tulsa World*, October 31, November 1, 1919; *New Solidarity*, November 8, 1919.

31. *Tulsa World*, November 2, 4, 5, 1919; *New Solidarity*, November 8, 1919; Lyons, *Assignment in Utopia*, 17-18.

32. *Tulsa World*, November 6, 7-9, 1919; *New Solidarity*, November 15, 29, 1919.

33. Lyons, writing eighteen years later, also said the jury vote was eleven-to-one for conviction. This clearly is incorrect. Lyons, *Assignment in Utopia*, 20; *Tulsa World*, November 11, 14, 1919; *New Solidarity*, November 15, 1919.

34. Moore to General Defense Committee of the IWW, April 14, 1920, published in *Solidarity*, April 24, 1920; see also *Solidarity*, May 22, 29, 1920.

35. Moore had asked Cole to quash the special venire because the jury commission had been appointed in violation of state law, but Cole refused. *Tulsa World*, May 14, 20, 21, 25-30, 1920; *Solidarity*, May 29, June 19, 1920.

36. *Tulsa World*, June 4-6, 8, 1920; *Solidarity*, ibid.

37. The union owed Fred Moore $1,000 and $2,689 for trial costs, as well as $318 for transcripts. *Solidarity*, June 12, 26, July 3, 1920; *Tulsa World*, June 9, 1920.

38. While Brissenden and Bohn saw the Australian laws as precursors of criminal syndicalism, Eldridge Dowell found the roots of the laws in the Alien and Sedition Acts of 1798. For the text of the bill, see Oklahoma S.B. 242, chap. 70, *Acts of the Seventh Legislature*, 110-12. For sabotage defined, see sec. 2. Eldridge Foster Dowell, *A History of Criminal Syndicalism Legislation in the United States*, Johns Hopkins University Studies in Historical and Political Science, ser. 56 no. 1 (Baltimore: Johns Hopkins Univ. Press, 1939), 13-15, 17-18, 18n, 24-25, 139; Brissenden, *The IWW*, 282n; William E. Bohn, "International Notes," *International Socialist Review* 17 (April 1917): 629.

39. Historian Paul Murphy notes that while Justice Brandeis believed the federal government was obligated to protect minority groups—as exemplified in his dissent in *Gilbert v. Minnesota*, 254 U.S. 325 (1920), a case involving repression of the Non-Partisan League—his opinion was not shared by other justices. Not until the case of *Gitlow v. New York*, 268 U.S. 562, in 1925 would the Court's position shift to Brandeis's view. Paul L. Murphy, *World War I and the Origins of Civil Liberties in the United States* (New York: W. W. Norton, 1979), 268.

40. Dowell, *Criminal Syndicalism Legislation*, 52n.

41. Ibid.; Smith, "Criminal Syndicalism in Oklahoma," 17-18, 47n; *Oklahoma Leader*, June 20, 1918; *Otter Valley Socialist*, May 10, 1916.

42. Dowell, *Criminal Syndicalism legislation*, 19, 19n, 66-67, 68n, 82.

43. Ibid., 71-73; Robertson to Harrison, February 14, 1919, in Box 24, Folder 10, Robertson Papers, OSA.

44. "The Anti-Syndicalist Laws," *One Big Union Monthly* 1 (April 1919): 9.

45. According to F. Thompson and Murfin, the union had an income of $77,968, indicating a membership of about 35,000. But its legal fees amounted to $101,809 for attorneys' fees, $8,985 for witnesses' expenses, and $29,603 for relief payments to prisoners and their families. This suggests the drive may also have been an attempt, however desperate, simply to raise funds to pay off debts. F. Thompson and Murfin, *First*

Seventy Years, 129-31; "Big Drive Is on in the Harvest Fields," *Rebel Worker,* June 15, 1919; "Agricultural Workers, Attention!" *One Big Union Monthly* 1 (April 1919): 53; *New Solidarity,* January 4, 1919.

46. "Agricultural Workers, Attention!" *One Big Union Monthly* 1 (April 1919): 53; McWilliams, *Ill Fares the Land,* 98-99; D. N. Simpson and Mat K. Fox, "Agricultural Workers Industrial Union No. 110 Bulletin No. 32, May 5, 1919," *One Big Union Monthly* 1 (June 1919): 57; "Agricultural Workers Industrial Union No. 110 Bulletin No. 37, June 9, 1919," *One Big Union Monthly* 1 (July 1919): 57; General Harvest Information Letter, 1919, and U.S. Department of Labor Circular no. 8, Kansas City, May 22, 1919; both in Box 26, Folder 10, Robertson Papers, OSA; *New Solidarity,* May 24, 31, 1919.

47. George Aldridge noted that many of the men arrested with him on June 11 were veterans. Two had been out of the army just three days. "The soldiers are beginning to see they will have to fight for democracy here in the harvest fields or it will all stay in France," he wrote. Aldridge to editor, *New Solidarity,* June 20, 1919; *New Solidarity,* May 31, June 28, 1919; "Many Agricultural Workers in Prison," *Rebel Worker,* July 1, 1919; "Crises Near in Harvest Fields," *Rebel Worker,* July 15, 1919; "The Campaign of the Agricultural Workers," *One Big Union Monthly* 1 (August 1919): 8; *Enid Daily Eagle,* January 18, 19, 1920; McWilliams, *Ill Fares the Land,* 98-99.

48. The newspaper accounts contain several errors. The *Enid Daily News* said he joined IWW in Missouri in August 1917, while the *Enid Daily Eagle* claimed he joined in August 1918. On this and other points the transcript should be considered more reliable. On Terrell's election to the GOC at the AWIU spring conference at Sioux City, Iowa, on April 23, 1919, and other matters, see direct testimony of Grover Jackson Terrell, Transcript, *State of Oklahoma v. Grover Jackson Terrell,* Criminal Case No. 520, County District Court, Enid, Oklahoma, 471, 480-85, 602-603, 613-16, 639. A copy of the transcript is in Criminal Court of Appeals Records, Oklahoma State Archives, Department of Libraries, Oklahoma City (hereafter Transcript, *State v. Terrell*); *Enid Daily Eagle,* January 18, 1920; *Enid Daily News,* January 18, 1920.

49. The indictment is dated September 10, 1919, but it is somewhat unclear why a new indictment was required. Moore and Bonstein, however, seemed to have missed the hearing on this new information. See Aldridge to Mat Fox and Fox to Moore, undated telegrams, in *New Solidarity,* September 30, 1919; ibid., July 5, 12, 26, August 2, 9, 30, 1919; Transcript, *State v. Terrell,* 30, 38, 74, 471.

50. Interestingly enough, Terrell and Aldridge both had a chance to escape when several prisoners broke out of the Enid jail in July 1919. C. W. Anderson, writing to *New Solidarity*, said, "They had not budged when the other birds flew because they had nothing to run away for," *New Solidarity*, July 26, 1919; Transcript, *State v. Terrell*, 212-469; *Enid Daily News*, January 17, 1920.

51. *Enid Daily News*, January 17, 1920; *Enid Morning News*, January 18, 1920; *Enid Daily Eagle*, January 18, 1920.

52. Much of the objections also involved rather heated arguments over legal minutiae between Moore and assistant prosecutor Robert E. Smith, which one reporter said became comical. Transcript, *State v. Terrell*, 485, 492, 504, 511-92.

53. Transcript, *State v. Terrell*, 618, 623, 635-36; *Enid Daily Eagle*, January 18, 1920; *Enid Daily News*, January 18, 1920; "Terrell Declares Self for Red Flag," *Enid Daily News*, January 19, 1920.

54. *Enid Daily Eagle*, January 19, 1920; *New Solidarity*, February 28, 1919; *Industrial Pioneer* 1 (May 1921): 56.

55. Aldridge, however, would have future legal problems in Oklahoma. He and three other Wobblies were arrested in Medford in the summer of 1920 and accused of the murder of a railroad policeman. The charge proved false, and Aldridge was released. *Solidarity*, July 31, August 21, 1920; *Industrial Solidarity*, September 17, 1921 (title changed from *Solidarity* with this issue); *Enid Daily News*, January 20, 1920; *Industrial Pioneer* 1 (May 1921): 56.

56. Terrell was sentenced in Los Angeles County court on July 12, 1923, to serve one to fourteen years. See San Quentin warden J. A. Johnston to Garfield County Sheriff O. A. Lincoln, August 7, 1923, copy in appendix, Transcript, *State v. Terrell*; on the appellate court's action, see *Terrell v. State of Oklahoma* 217 *Pacific Reporter* 900.

57. John Higham, *Strangers in the Land: Patterns of American Nativism, 1860-1925* (New York: Atheneum, 1969), 227

58. Murphy, *World War I and Civil Liberties*, 86.

59. Beckham County reached only 17 percent of its quota for Liberty Bonds. J. Thompson, *Closing the Frontier*, 189.

60. The quotation referred specifically to the indictment for opposing the draft of A. D. Engel, secretary of the Chattanooga, Oklahoma, SP local. *Harlow's Weekly*, August 1, 1917; Higham, *Strangers in the Land*, 222-23.

Chapter 7. A Brief Renaissance

1. That the IWW was aware of these changes can be seen in Nick Wells, "As to Lubrication," *Industrial Pioneer* 1 (November 1921): 22-23,

and "Coming Fuel and Transportation Trust," *Industrial Pioneer,* 2d ser., 1 (October 1923): 21; Robert H. Zieger, *American Workers, American Unions, 1920-1985,* (Baltimore: Johns Hopkins Univ. Press, 1986), 6-7; James R. Mock and Evangeline Thurber, *Report on Demobilization* (Norman: Univ. of Oklahoma Press, 1944), 126-44; Noggle, *Into the Twenties,* 10; Solberg, *Oil Power,* 73-74, 80-82, 108-109, 155-60; Dubofsky, *We Shall Be All,* 473.

2. John S. Gambs, *The Decline of the IWW.* New York: Russell and Russell, 1966; Dubofsky, *We Shall Be All,* 473.

3. Dubofsky notes the late Fred Thompson, the IWW's official historian, recalled 1919 with relish and apparently to his dying day believed the revolution was just around the corner. But Dubofsky also claimed the 1919 convention showed no links to the past because the delegates were unfamiliar, which was clearly not true. Many attendees were veterans of earlier campaigns who had not previously taken major leadership roles. *We Shall Be All,* 453.

4. An unidentified clipping in the Cole Collection, Box 37, Folder 2, dated June 15, 1918, refers to the use of the German POWs. *Harlow's Weekly,* January 23, June 12, 1918; *Oklahoma Farmer-Stockman,* May 25, June 25, July 10, July 25, 1918; Fast, "Agricultural Workers Organization," 109.

5. Mat K. Fox, "The Story of No. 400," *One Big Union Monthly* 1 (September 1919): 49, "Agricultural Workers Industrial Union No. 400 Bulletin No. 29, March 31, 1919," *One Big Union Monthly* 1 (May 1919): 56; Mat K. Fox and Walter Sheridan, "Agricultural Workers Industrial Union No. 400, IWW" *One Big Union Monthly* 1 (March 1919): 58; *New York Times,* June 14, 1919; *Solidarity,* May 24, 1919.

6. One of the men arrested in Bartlesville, an oil worker named George Schwartz, seems to be the only Wobbly jailed in Oklahoma who faced a warrant for deportation. For other arrests, see *New Solidarity,* February 15, March 1, May 17, July 5, 12, 26, August 2, 30, 1919; *Oklahoma Leader,* March 15, 1919; D. N. Simpson and Mat K. Fox, "Agricultural Workers Industrial Union No. 110 Bulletin No. 45, August 11, 1919," *One Big Union Monthly* 1 (September 1919): 50.

7. About 50 percent of the hands in 1919 were former soldiers. Jamieson, *Labor Unionism,* 404-405; Fast, "Agricultural Workers Organization," 115; F. Thompson and Murfin, *First Seventy Years,* 130; Simpson and Fox, "Agricultural Workers Industrial Union No. 110 Bulletin No. 56, November 10, 1919," *One Big Union Monthly* 1 (December 1919): 53-54.

8. Jamieson, *Labor Unionism,* 404-405; *Tulsa World,* May 16, 1920; *Daily Oklahoman,* May 25, 1920; Oklahoma State Department of Labor, *Annual*

Report for 1932 (Oklahoma City: OSDL, 1932), *Bulletin 10-A* (Oklahoma City: OSDL, 1928), 32-33. Copies in Oklahoma State Archives, Department of Libraries, Oklahoma City; *Solidarity*, June 26, August 21, 1920.

9. *Solidarity*, October 9, 16, 1920.

10. Officials had estimated a need for forty to fifty thousand hands, clearly an exaggeration, but it does reveal that the overplanted harvest required additional help. Some state newspapers even urged that Oklahoma recruit hands from New York's reported 486,000 unemployed. Donald E. Green, "Beginnings of Wheat Culture in Oklahoma" in *Rural Oklahoma*, ed. Donald E. Green (Oklahoma City: Oklahoma Historical Society, 1977), 67-68; Hale, "People of Oklahoma," 57; *Daily Oklahoman*, June 14, 15, 1921; Oklahoma State Department of Labor, *Annual Report for 1932*, Bulletin 10-A; Oklahoma State Department of Labor, *Bulletin No. 2* (November 1, 1921), 1; *Industrial Solidarity*, November 25, 1922.

11. Writing in the June 25 issue of *Solidarity*, Mexican correspondent Jose F. Montemayor said before the harvest started six Mexicans had already been deported: four from Oklahoma City and two from Lawton. Oklahoma State Department of Labor, *Oklahoma Department of Labor Bureau of Free Employment Bulletin No. 6* (Oklahoma City: OSDL, 1922), 34; *Solidarity*, May 28, June 11, 18, 25, July 9, 1921; "General Defense Committee Bulletin," *Industrial Pioneer* 1 (August 1921); Harry Feinberg, "Defense News," *Industrial Pioneer* 1 (August 1921): 60; Lescohier, "Harvesters and Hoboes," 503-504.

12. *Solidarity*, July 16, 1921; William Dimmit, "An Organized Harvest," *Industrial Pioneer* 1 (September 1921): 3-5.

13. *Enid Daily Eagle*, June 15, 16, 1922; *Industrial Solidarity*, May 27, June 3, 10, 1922.

14. *Industrial Solidarity*, June 24, July 1, 1922.

15. Initially *Industrial Solidarity* reported fifty-six men arrested, but later reports and the *Enid Daily Eagle* agree on fifty-seven. The arrests apparently followed sensationalized reports of police and Wobbly clashes in the Harvest Belt printed in Oklahoma City and Wichita papers. *Industrial Solidarity*, June 17, 24, July 1, 1922; *Enid Daily Eagle*, June 12, 1922.

16. *Enid Daily Eagle*, June 12, 14, 1922; *Industrial Solidarity*, July 1, 1922.

17. "Don't Rest on Enid. Push on 110! Now for a Clean Finish," *Industrial Solidarity*, July 1, 1922; *Industrial Solidarity*, June 24, 1922; *Enid Daily Eagle*, June 15, 16, 1922.

18. The actual incident involved only ten men, but some out-of-state press reports claimed as many as fifty people were involved. One AWIU member said the jury's verdict may have been influenced by a Ku Klux Klan parade and rally before the inquest. *Industrial Solidarity*, June 24,

NOTES TO PAGES 150-54   247

July 1, 1922; *Enid Daily Eagle*, June 18, 19, 20, 22, 1922; Investigator W. C. Gordon to State Atty. Gen. George Short, June 29, 1922, Box 28, Folder 2082, AG Files, OSA.

19. *Industrial Pioneer*, 2d ser., 1 (July 1923): 3-4.

20. "Wobblies Are Invading City," *Daily Oklahoman*, May 20, 1923; *Minutes of the Fourteenth Convention of the Agricultural Workers Industrial Union No. 110, May 20-24, 1923, Oklahoma City* (Chicago: Printing and Publishing Workers Industrial Union no. 450, 1923), sixteen page pamphlet in Box 44, Folder 5, IWW Collection, ALUA/WSU, 2, 4-5.

21. *Minutes of Fourteenth AWIU Convention*, 12-14; *Industrial Pioneer*, 2d ser., 1 (July 1923): 3-4.

22. *Minutes of Fourteenth AWIU Convention*, 15.

23. Ibid., 16; *Oklahoma City Times*, May 21, 1923; "The Law Is Sufficient," *Daily Oklahoman*, May 27, 1923.

24. *Industrial Solidarity*, May 26, June 16, 30, 1923; *Daily Oklahoman*, May 25, 27, 28, June 2, 1923; Oklahoma State Department of Labor, *Bulletin No. 10*, 8, 17-18; *Annual Report for 1932; Bulletin 10-A*, 32-33.

25. *Daily Oklahoman*, May 26, June 1, 3, 1923; *Industrial Solidarity*, June 23, 1923; *Industrial Pioneer*, 2d ser., 1 (July 1923): 5.

26. *Industrial Solidarity*, June 23, 30, 1923.

27. The AWIU also passed resolutions formally condemning violence and proposing means to counter the "capitalist" press image of the IWW. *Industrial Pioneer*, 2d ser., 1 (July 1923); F. Thompson and Murfin, *First Seventy Years*, 149; *Minutes of the Fifteenth Convention of the Agricultural Workers Industrial Union No. 110, October 8-18, 1923, Fargo, North Dakota* (Chicago: Printing and Publishing Workers Industrial Union No. 450, 1923), in Box 44, Folder 6, IWW Collection, ALUA/WSU 20, 22, 28.

28. *Minutes of the Fifteenth Convention of the AWIU*, 6; *Industrial Solidarity*, April 26, 1924.

29. J. Thompson, *Closing the Frontier*, 169, 174.

30. The special oil workers issues were *Industrial Solidarity*, January 27, 1923, and *Industrial Pioneer*, 2 (November 1924); "Coming Fuel and Transportation Trust," 21; Albert Barr, "Oil and Oil Workers," *Industrial Pioneer* 1 (September 1921): 11-12.

31. Many of the articles for oil workers consisted solely of job listings and notes urging Wobblies to take the jobs. Other cities where oil workers were detained included Ardmore and Walters. Fox and Sheridan, "AWIU No. 400, IWW," 59; Simpson and Fox, "AWIU Bulletin No. 56," 54. *New Solidarity*, January 11, February 8, 15, March 1, 1919, February 14, 28, March 20, 1920.

32. Standard first introduced "company unions" in the oil industry on April 1, 1918, at its Bayonne, New Jersey, refinery, the scene of two violent strikes in July 1915 and October 1916. The Bayonne "union" was modeled on the one introduced in Colorado following the Ludlow massacre. Some Oklahoma companies adopted the employee representation plan early, but most did not do so until the 1930s, and then mostly as a means of subverting Section 7a of the National Industrial Recovery Act. Oklahoma State Department of Labor, *Bulletin No. 2*, 1; O'Connor, *OWIU-CIO History*, 20-21, 95-98; Daniel Horowitz, *Labor Relations in the Petroleum Industry*, 2d ed. (New York: Works Progress Administration, 1938), 28-37, 40.

33. O'Connor, *OWIU-CIO History*, 9-13, 18-24; Horowitz, 63-65; Nick Wells, "Tactics in Oil," *Industrial Pioneer* 1 (December 1921): 26. See also Wells, "As to Lubrication," 22-23.

34. William Dimmit, "Splitting the Big Drive," *Industrial Pioneer* 1 (December 1921): 11; *Solidarity*, July 2, 9, September 10, 1921.

35. The GOC was composed of Robert Smith, William Danton, John Obrein, M. H. McDonald, and Harvey Karnston. Apparently neither minutes nor any other record exists of that first OWIU meeting in Oklahoma City. *Solidarity*, August 28, October 16, 1920; *Minutes of the Twelfth Convention of AWIU No. 110, October 17-25, 1921, Omaha, Nebraska* (Chicago: Printing and Publishing Workers Industrual Union No. 450, 1921), in Box 44, Folder 3, IWW Collection, ALUA/WSU, 2, 9, 24.

36. For a complete discussion of the strike, see Nigel Sellars, "Butchers and Businessmen: The 1921 Packinghouse Strike in Oklahoma City," (presented at the annual meeting of the Southwest Social Sciences Association, March 29, 1990). A history of the Open Shop Division and the Employers' Association is in Reinhart, "Open Shop Movement in Oklahoma"; *Daily Oklahoman*, April 27, 1919.

37. *Industrial Solidarity*, January 7, 14, 21, 1922.

38. The *Oklahoma News* headlined was "Lynched by the I.W.W.?" Hughes later denied accusing the Wobblies, saying that IWW members were safe as long as they obeyed the laws and that he had no plans to raid their headquarters. The Amalgamated's leaders called off the strike January 31, but Oklahoma City strikers voted overwhelmingly to stay out. *Oklahoma News,* January 18, 1922; *Oklahoma Leader,* January 12, 19, 30, February 2, 1922; *Industrial Solidarity*, January 28, February 18, 1922.

39. Henry Bradley to editor, November 10, 1921, in *Industrial Solidarity*, November 12, 1921; *Industrial Solidarity*, March 11, 1922; "Fellow Workers," Oil Workers Industrial Union no. 230, 1921, Oklahoma City, photocopy

NOTES TO PAGES 158-63   249

in author's possession; "Oil Workers!" (Oklahoma City: Oil Workers Industrial Union no. 230, n.d. [1922]), in Box 169, IWW Collection, ALUA/WSU.

40. *Industrial Solidarity*, May 27, June 3, 1922.

41. *Industrial Solidarity*, April 22, July 15, 1922.

42. "Organizing the Oil Industry No. 5—the Tank Builders," *Industrial Solidarity*, August 6, 1921; "The Tankie," undated interview by Ned DeWitt, in Kenny Franks, ed., *Voices from the Oil Fields* (Norman: Univ. of Oklahoma Press, 1989), 175, 180-81.

43. "Organizing the Oil Industry No. 5," ibid.

44. "Piece Work and the Tank Builder" (Chicago: Industrial Workers of the World, 1923), four-page pamphlet, copy in Box 170, Pamphlet File, IWW Collection, ALUA/WSU. The same material also appeared in *Industrial Solidarity*, January 27, 1923, where it was attributed to union member "WP-13."

45. DeNoya was also known by the more common nickname of "Whizzbang." John W. Morris, *Ghost Towns of Oklahoma* (Norman: Univ. of Oklahoma Press, 1978), 64-65; Edward E. Anderson, "Oklahoma Tank Builders Win Two Strikes," *Industrial Solidarity*, December 9, 1922; "Tank Builders Strike Again at Whizzbang," *Industrial Solidarity*, January 13, 1923.

46. Anderson, "Oil Workers Come Back"; *Industrial Solidarity*, October 28, 1922; Horowitz, *Labor Relations*, 63-65.

47. Dubofsky, *We Shall Be All*, 453; Gambs, *Decline of the IWW*; Conlin, *Bread and Roses, Too*, 141-44.

48. Saposs, *Left-Wing Unionism*, 142-43.

49. William Preston rightly notes that industrial unionism had many seeds, that the IWW as a whole was smaller than any AFL-affiliated industrial union, and that the Wobblies failed to organize the mass production industries that formed the core of the CIO. But the petroleum industry is the exception, and the OWIU's influence among oil workers should not be underestimated. Preston, "Shall This Be All?" 448-49.

Chapter 8. Decline and Fall, 1922-1930

1. *Industrial Solidarity*, March 4, 1922.

2. That the IWW understood the effects of the postwar technological advances can be seen in the following: Joseph Ostrander, "The New Labor Displacers," *Industrial Pioneer*, 2d ser., 2 (July 1924): 31-32, 34; "Farm Machinery Displacing Workers," *Industrial Solidarity*, July 28, 1926. Also, as Robert L. Tyler has pointed out, the union's creation of an

education bureau after the war also shows the IWW recognized the need to harness technological change as part of the revolution. See Robert L. Tyler, "The IWW and the Brainworkers," *American Quarterly* 15 (spring 1963): 51. Quotation is from *Industrial Pioneer*, 1st ser. (July 1923): 5.

3. For the Klan as a continuation of the martial spirit of the war, see Higham, *Strangers in the Land*, 222-23 and Noggle, *Into the Twenties*, 100-101; Blake's speech, "The K. K. Kraze," is cited in Nina A. Steers, "The Ku Klux Klan in Oklahoma in the 1920s" (master's thesis, Columbia University, 1965), 48; Jensen, *Price of Vigilance*, 289; Kenneth T. Jackson, *The Ku Klux Klan in the City, 1915-1930* (New York: Oxford Univ. Press, 1967), 237.

4. Steers, "Ku Klux Klan," 69, 133, 65-66, 90, 98.

5. Jackson, *Ku Klux Klan*, 84-85, 239; Scott Ellsworth, *Death in a Promised Land: The Tulsa Race Riot of 1921* (Baton Rouge: Louisiana State Univ. Press, 1982), 20-22; *Oil and Gas Journal*, August 26, 1921.

6. *Imperial Nighthawk*, August 8, 1923.

7. F. Thompson and Murfin, *First Seventy Years*, 148; *Industrial Solidarity*, January 21, February 4, 1922.

8. *Industrial Solidarity*, July 22, 1922, May 19, 1923; David M. Chalmers, *Hooded Americanism: The History of the Ku Klux Klan* (New York: New Viewpoints, 1981), 51.

9. "KKK Businessmen Starving Oklahoma," *Industrial Solidarity*, April 5, 1924. See also *Industrial Solidarity*, January 21, February 11, 1922, January 6, 1923.

10. California tried 317 IWW members, of whom 140 were convicted. Their sentences ranged from one to fourteen years, with an average of four years. F. Thompson and Murfin, *First Seventy Years*, 149.

11. *Tulsa Tribune*, July 23, 27-31, 1922; William D. McBee, *The Oklahoma Revolution* (Oklahoma City: Modern Publishers, 1956), 2. McBee was speaker of the Oklahoma House during Walton's impeachment. Alexander, "Invisible Empire," 159; *New York Times*, May 28, 1923.

12. Interview with John J. Carney, in J. M. Smith, "Criminal Syndicalism in Oklahoma," 57; *Oklahoma Leader*, August 19, September 28, 1923.

13. *McAlester News-Capital*, December 28, 1922; "Trial record and transcript of *State of Oklahoma v. Arthur Berg*, Criminal Case No. 2146, Pittsburg County, with filings for writ of error," in Box 127, Folder 10, IWW Collection, ALUA/WSU, 82-83, 93, 100, 158. Another copy is in the Criminal Court of Appeals Records, Oklahoma State Archives, Department of Libraries, Oklahoma City (hereafter Transcript, *State v. Berg*).

NOTES TO PAGES 168-73   251

14. *McAlester News-Capital*, January 17, 1923.
15. "John J. Carney," in Luther B. Hill, ed., *History of the State of Oklahoma*, vol. 2 (Chicago: Lewis Publishing Co., 1908), 399; *Daily Oklahoman*, January 27, March 17, 1907; "John Carney, Pioneer State Lawyer, Dies," *Daily Oklahoman*, January 16, 1954; *Lexington Leader*, September 18, 1908; *Otter Valley Socialist*, January 5, 1915.
16. Transcript, *State v. Berg*, 7, 12, 14-16, 18-19, 20.
17. Ibid., 85-87, 105, 109-11, 121, 123-24.
18. Ibid., 23, 84.
19. Ibid., 82-83, 130-31, 143-44, 147; *Industrial Solidarity*, August 20, 1923.
20. Transcript, *State v. Berg*, 20, 78-79, 122.
21. *McAlester News-Capital*, February 28, 1923; *Pittsburg County Guardian*, March 1, 1923; *Industrial Solidarity*, August 20, 1923.
22. George G. Suggs, Jr., *Union Busting in the Tri-State: The Oklahoma, Kansas, and Missouri Metal Workers Strike of 1935* (Norman: Univ. of Oklahoma Press, 1986), 7, 11.
23. The *Miami (Oklahoma) Daily Record-Herald*, May 31, 1923, notes that financier Charles Schwab owned the Quapaw Mining Corporation, one of the largest producers in the district. Suggs, *Union Busting*, 8, 10-12, 22; *Oklahoma Leader*, August 19, September 28, 1923; T. Lane Carter, "Economic Conditions in the Joplin District," *Engineering and Mining Journal* 90 (October 15, 1910): 760; Mills, "Joplin Zinc," 658.
24. Suggs, *Union Busting*, 9; *Oklahoma Labor Unit*, February 28, 1914; Miner "Alex Mac———," quoted in Mills, "Joplin Zinc," 658.
25. *Homer Wear v. State of Oklahoma*, transcript and brief of plaintiff in error, March 5, 1925, copy in Criminal Court of Appeals Records, Oklahoma State Archives, Department of Libraries, Oklahoma City, 28-30 (Hereafter Transcript, *Wear v. State*); *Industrial Solidarity*, June 23, 1923.
26. *Miami (Oklahoma) Daily Record-Herald*, July 11, August 7, 1923; Transcript, *Wear v. State*, 44, 103-105; *Industrial Solidarity*, August 20, 1923.
27. *Industrial Solidarity*, September 1, 1923.
28. *Industrial Solidarity*, August 20, 1923; *Miami (Oklahoma) Daily Record-Herald*, September 26-27, 1923; Transcript, *Wear v. State*, 2, 9, 11, 20, 24-25.
29. *Berg v. State*, 29 *Oklahoma Criminal Court of Appeals* 112, 233 Pacific Reporter 497; *Industrial Solidarity*, February 4, 11, 1925.
30. *Berg v. State*, ibid; *Wear v. State*, 30 *Oklahoma Criminal Court of Appeals* 118, 235 Pacific Reporter 271; *Industrial Solidarity*, February 4, 11, 18, 25,

March 4, May 13, 27, June 4, 1925; *Industrial Unionist,* May 2, June 6, 1925; *General Office Bulletin,* May 1924, 23, mimeo in Box 31, Folder 24, IWW Collection, ALUA/WSU.

31. *Industrial Solidarity,* February 4, 11, 18, 25, March 4, May 13, 27, June 4, 1925;; *Industrial Unionist,* May 2, June 6, 1925.

32. F. Thompson and Murfin, *First Seventy Years,* 150-51; Taft, "IWW in the Grain Belt," 64; Dubofsky, *We Shall Be All,* 466-67; Gambs, *Decline of the IWW,* 99-100.

33. Dubofsky, *We Shall Be All,* 457, 461-64; Gambs, *Decline of the IWW,* 106-15.

34. F. Thompson and Murfin, *First Seventy Years,* 150-51; Dubofsky, *We Shall Be All,* 458-59; Gambs, *Decline of the IWW,* 101, 104-105; "Why Eleven Members of the IWW Imprisoned at Leavenworth Refused Conditional Pardons," two-page broadsheet dated July 30, 1923, printed at Philadelphia by H. F. Kane, Copy in Alfred and Rose Anderson Collection, ALUA/WSU.

35. Gambs, *Decline of the IWW,* 101, 104-105; Dubofsky, *We Shall Be All,* 465-66.

36. Dubofsky, *We Shall Be All,* 465-66; Gambs, *Decline of the IWW,* 99-100, 110-11; F. Thompson and Murfin, *First Seventy Years,* 155n; *Industrial Solidarity,* October 22, November 5, 19, December 3, 1924. Minutes of the convention are reprinted in several issues.

37. The four secessionist unions were those of the lumberjacks, the construction workers, the metalworkers, and the railroad workers. Gambs, *Decline of the IWW,* 118, 123.

38. Ball, *Fascinating Oil Business,* 180-82; *Daily Oklahoman,* June 3, 1923.

39. Ball, *Fascinating Oil Business,* 180-85; Solberg, *Oil Power,* 121-22; Hale, "People of Oklahoma," 59.

40. "Financial Statement of OWIU No. 230, Oklahoma City, September 1923," two mimeographed pages and *OWIU Bulletin,* September 30, 1923, two mimeographed pages, both documents in Box 51, Folder 7, IWW Collection, ALUA/WSU. "Minutes of OWIU Convention, Oklahoma City, October 16-18, 1923," sixteen legal-sized mimeographed pages, copy in Box 51, Folder 3, IWW Collection, ALUA/WSU, 3-5; *Minutes of the Fourteenth AWIU Convention,* 14.

41. "To the members of the IWW," one-page, undated flyer published in New York by Bert Lorton and the anti-Chicago faction; copy in Alfred and Rose Anderson Collection, Ser. 1, Box 3, Folder 15, ALUA/WSU. Gambs, *Decline of the IWW,* suggests the flyer was published in summer 1923, 104n; *Minutes of OWIU Convention,* 6, 9-10.

42. Dubofsky, *We Shall Be All*, 448; Ralph Colescott, delegate no. 40-54, to General Executive Board, *General Office Bulletin*, March 1924, copy in Box 31, Folder 22, IWW Collection, ALUA/WSU, 27-28.

43. *General Office Bulletin*, March 1924, 29, and February 1924, 27; both in Box 31, Folder 21, IWW Collection, ALUA/WSU; *Industrial Solidarity*, April 5, 1924.

44. *Industrial Solidarity*, April 26, 1924.

45. According to the "Pie-Card Gang" flyer, a "gang" led by Tom Doyle, the general secretary-treasurer, seized the IWW headquarters and the offices of the Railroad Workers Industrial Union. The armed "gang" included Joe Fisher, Frank Gallagher, Forrest Edwards, and E. W. Latchem. The action had been approved by the heads of the AWIU and OWIU. "Pie-Card Gang Raids I.W.W. Headquarters with Guns and Drives the GEB into the Street," one-page flyer, n.p., n.d., but clearly published after July 29, 1924 by supporters of James Rowan. Copy in vertical files, IWW Collection, in Labadie Collection, University of Michigan, Ann Arbor; *General Office Bulletin*, March 1924, 22; *OWIU Bulletin*, October 31, 1923, in Box 51, Folder 8, IWW Collection, ALUA/WSU; *Industrial Solidarity*, April 12, September 29, October 13, 22, 29, 1924.

46. *Industrial Unionist*, December 16, 1925.

47. *Industrial Solidarity*, September 30, November 15, December 2, 1925; March 10, April 14, June 23, July 1, 1926.

48. *Industrial Solidarity*, May 13, 1925; *Industrial Pioneer*, 2d ser., 1 (June 1923): 45; and *Industrial Pioneer* 2d ser., 2 (October 1924): 27-28; Harder, "Wheat Production," 72; D. E. Green, "Beginnings of Wheat Culture," 73; Hale, "People of Oklahoma," 57.

49. Isern, *Bull-Threshers*, 188; Isern, *Custom Combining*, 19; Jamieson, *Labor Unionism*, 405; *Industrial Solidarity*, July 21, August 11, 1926; Hamilton Wright, "Harvesting the West's Great War Crops," *Forum* 58 (October 1917): 483.

50. Oklahoma State Department of Labor, *Bulletin No. 10-A*, 10-12; Oklahoma State Department of Labor, *Annual Report for 1932*, 32-33; F. Thompson and Murfin, *First Seventy Years*, 149.

51. In 1925, an Enid reporter interviewed Denny Taylor, a sixty-year-old Ohio man who had made the harvest for "many years," the article said. While probably not a Wobbly, Taylor's sentiments reflect the feelings of many harvest stiffs. "Old Time Harvest Worker Disgusted with Use of Revolutionized Methods," *Enid Morning News*, June 6, 1925; F. Thompson and Murfin, *First Seventy Years*, 95; John J. Hader, "Honk

Honk Hobo," *Survey* 60 (August 1, 1928): 453-54; Oklahoma State Department of Labor, *Bulletin No. 10*, 8, 17-18; McWilliams, *Ill Fares the Land*, 101.

52. Hader, *Honk Honk Hobo*, 453-54, quotation on 455.

53. *Industrial Solidarity*, June 3, 10, 17, July 9, 1925, May 12, June 2, 1926; *Enid Morning News*, June 6, 1925.

54. *Industrial Solidarity*, May 12, June 2, 23, 1926.

55. *Industrial Solidarity*, May 4, June 1, 22, November 23, 1927, May 22, June 12, 26, 1929, May 26, June 16, 1931; *Minutes of the Twentieth Convention of Agricultural Workers Industrial Union 110, Williston, North Dakota, October 10, 1928* (Minneapolis: Allied Printing Trades Council, 1928), copy in Box 44, Folder 13, IWW Collection, ALUA/WSU, 11; "Minutes of the Spring Conference of the AWIU No. 110, Omaha, Nebraska, June 1, 1933," four-page typescript in Box 44, Folder 17, IWW Collection, ALUA/WSU.

56. *Industrial Solidarity*, January 7, February 4, April 8, 1925; June 23, July 21, August 11, 18, September 16, November 3, 1926; report of A. S. Embree, dated January 1, 1927, in "Minutes of Special Session of GEB, January 3-7, 1927, Chicago, Ill.," copy in Box 7, Folder 13, IWW Collection, ALUA/WSU; Joseph Wagner, *Report of the General Secretary to the Twenty-First Convention of the Industrial Workers of the World, November 12-18, 1934*, 7, copy in Box 7, Folder 26, IWW Collection, ALUA/WSU.

57. Montgomery, *Workers' Control*, 100; J. Green, *World of the Worker*, 101.

58. Dubofsky, *We Shall Be All*, 466.

59. Ostrander, "The New Labor Displacers," *Industrial Pioneer*, 2d ser., 2 (July 1924): 31-32, 34; "Farm Machinery Displacing Workers," *Industrial Solidarity* (July 28, 1926); Trachtenberg, *Incorporation of America*, 21-22.

Epilogue

1. E. W. Latchem, "Some Worthwhile Reflections about the Changing Times," fourteen-page typescript, undated, copy in Box 145, Folder 30, IWW Collection, ALUA/WSU, 2-3, 7-8.

2. Latchem, "Some Worthwhile Reflections," 8-9; E. W. Latchem, "Introduction to 'The IWW in the West: Its Significance,'" eight-page typescript, with corrections, originally prepared for the annual meeting of the Western Historical Association, June 1965, Salt Lake City, Utah, carbon copy in Box 75, Folder 1, E. W. Latchem Papers, Small Processed Collections, ALUA/WSU, 2. Latchem joined the IWW in Calgary, Alberta, and worked with a railway union in British Columbia. He was

involved in migratory work through the Palmer Raids and later served as an IWW official.

3. *Tulsa Labor News*, June 1959.

4. Lyons, *Assignment in Utopia*, 20. Krieger's nemesis, J. Edgar Pew, went on to become chairman of Sun Oil of Philadelphia. See "Forty Years of the Oil Age" (New York: Petroleum Institute, 1944), 3; Holbrook, "Last of the Wobblies," 167

5. Earl Bruce White's work in tracking down the Wichita defendants through federal prison and Social Security records is highly useful. He is among the few researchers who have tried to trace rank-and-file Wobblies. Those with Oklahoma connections who became alcoholics with lengthy jail records included Arthur Blumberg and Peter Higgins. Both Paul Maihak and Joseph Gresbach were found dead—Gresbach in Jackson, Minnesota, in 1935, and Maihak in New York City in 1945. White gives no cause of death in either case, but one suspects their deaths were also alcohol-related. White, "Wichita Indictments," 285, 287-90; *Industrial Solidarity*, April 5, 1924. See also "Pie-Card" flyer.

6. White, "Wichita Indictments," 282-83.

7. *Industrial Worker*, (Chicago), August 22, 1925; Conlin, *Bread and Roses, Too*, 150.

8. F. Thompson and Murfin, *First Seventy Years*, 95-96, argue that the AWIU campaigns produced recruits who worked in other industries and who helped familiarize others with the union by circulating IWW literature at their jobs; "Joseph Murphy Interview," in Bird, Georgakas, and Shaffer, *Solidarity Forever*, 52; Karson, *American Labor Unions*, 210; Saposs, *Left-Wing Unionism*, 135-36, 148-49, 158. See also David Saposs to William Lieserson, February 28, 1919, quoted in Paul M. Buhle, "Marxism in the United States" (Ph.D. dissertation, University of Wisconsin-Madison, 1975), 153n.

9. Buhle argues that the IWW saw society as a great factory that treated workers as commodities and not as whole human beings. He contends this kept the IWW from bread-and-butter issues. This, however, seems to intellectualize IWW beliefs too much and fails to recognize that the Wobblies clearly did work for real-life issues. The organization was simply unable—structurally, financially, and tactically—to solidify those gains when they were achieved. Buhle, "Marxism in the United States," 44-45.

10. Applen, "Migratory Harvest Labor," 77; Donald Worster, *The Dust Bowl: The Southern Plains in the 1930s* (New York: Oxford Univ. Press, 1979), 58-59.

11. Jamieson, *Labor Unionism*, 266-76; for other references to WUW activity, see the exchanges between H. L. Mitchell of the STFU and J. W. Eakins of the WUW in the *Southern Tenant Farmers Union Papers, 1934-1970* (New York: Microfilming Corporation of America, 1971), esp. Mitchell to Eakins, January 11, February 7, 1936, and Eakins to Mitchell, January 13, 17, 22, 1936, on Reel 1; and Eakins to Mitchell, December 28, 1936, Reel 3.

12. Of the migrants who arrived in California between July 1, 1935, and March 31, 1938, 24 percent, or 58,153, were from Oklahoma, so it seems quite likely that a sizable number were involved with the CAWIU, which led a number of California strikes. The CAWIU later became the United Cannery, Agricultural, Packinghouse and Allied Workers of America and part of the CIO. See Paul Taylor, "Refugee Labor Migration to California, 1937," in Paul Taylor, *On the Ground in the Thirties* (Layton, Utah: Peregrine Smith, 1983), 222-23. The article originally appeared in *Monthly Labor Review* (April 1939); "The IWW and the Unemployed, 1913-1915," chap. 19 in Foner, *IWW History*; J. Green, *Grass-Roots Socialism*, 417, 418n; Jamieson, *Labor Unionism*, 265-66, 267. See also Patrick McGinnis, "'Share the Work': Ira M. Finley and the Veterans of Industry of America," in *Hard Times in Oklahoma: The Depression Years*, ed. Kenneth K. Hendrickson, Jr. (Oklahoma City: Oklahoma Historical Society, 1983), 22-46.

13. Davis, *Prisoners of the American Dream*, 53.

14. Angie Debo and John M. Oskison, eds., *Oklahoma: A Guide to the Sooner State* (Norman: Univ. of Oklahoma Press, 1941), 49.

15. Langston was publisher of the newspaper, while Sheldon served as editor. Purdy—who was secretary of the Farmer-Labor Reconstruction League from 1922 to 1924—and Fenton regularly wrote columns for the paper. *Oklahoma Labor*, November 22, 29, 1935; January 3, February 14, 21, April 2, 16, May 21, September 3, 1936; Victor S. Purdy, "Controversy of Craft versus Industrial Unionism," *Oklahoma Labor*, May 14, 1936.

16. "Walter Strong" was a pseudonym used by Ned DeWitt in his interview "What Do You Hear from Europe?" Oil in Oklahoma, WHC/OU; Hall, "Labor Struggles in the Deep South," 162, 165; O'Connor, *OWIU-CIO History*, 94-95, 162, 175.

17. There were also locals at Augusta and El Dorado, Kansas. O'Connor, *OWIU-CIO History*, 105, 162-64, 174-77, 180-83, 189-91, 217, 284-85, 346-49, 372-78; Horowitz, *Labor Relations*, 66-67, 72n, 69-72, 73-75; Walter Galenson, *The CIO Challenge to the AFL: A History of the*

American Labor Movement, 1935-1941 (Cambridge: Harvard Univ. Press, 1960), 409; *Oklahoma Labor*, October 1, 1936, March 4, 1937.

18. For a discussion of Oklahoma opposition to Section 7a and the National Labor Relations Act (Wagner Act), see Reinhart, "Open Shop Movement in Oklahoma," 36–37, 70–71; Dubofsky, *We Shall Be All*, 477–78.

19. Galenson, *CIO Challenge*, 409, 418–19; O'Connor, *OWIU-CIO History*, 346–49.

20. A sit-down strike also occurred in December 1936 at the Empire Oil and Refinery Company in Seminole, while Cushing Gasoline and Refining closed down before a strike could take place. See *Oklahoma Labor*, January 7, 1937; O'Connor, *OWIU-CIO History*, 346–49; Galenson, *CIO Challenge*, 418–19; J. M. Smith, "Criminal Syndicalism," 78–82.

21. J. M. Smith, "Criminal Syndicalism," 95–164, gives in-depth discussion of the "Bookseller" cases, as they were called. See also *Robert Wood v. State*, 77 Oklahoma Criminal Court of Appeals 305; *Ina Wood v. State*, 76 Oklahoma Criminal Court of Appeals 89; *Shaw v. State*, 76 Oklahoma Criminal Court of Appeals 271; *Jaffee v. State*, 76 Oklahoma Criminal Court of Appeals 95.

22. Besides the Oklahoma City papers, the *Norman Transcript* and *Tulsa Tribune* weighed in against the defendants and for the law, but the University of Oklahoma's student newspaper, the *Oklahoma Daily*, called the law an affront to the First Amendment. J. M. Smith, "Criminal Syndicalism," chap. 9, but esp. 133, 137, 138–39.

23. Patricia Cayo Sexton, *The War on Labor and the Left: Understanding America's Unique Conservatism* (Boulder, Colo.: Westview Press, 1991), 118–19.

24. Saposs, *Left-Wing Unionism*, 144–46. 152. 357.

25. Those who saw the "Bummery" as capturing the IWW include David Saposs, Joseph Conlin, and Paul Buhle.

26. Karson, *American Labor Unions*, 210.

27. Dave Archibald interview, Oral History Collection, Department of Libraries, Oklahoma City. Quoted in J. Green, *Grass-Roots Socialism*, 226n.

Bibliography

PRIMARY SOURCES

Manuscript Collections

Labadie Collection, University of Michigan, Ann Arbor
   IWW Collection
Oklahoma Historical Society, Oklahoma City
   Vertical Files—Labor, Socialist Party
Oklahoma State Archives, Department of Libraries, Oklahoma City
   Criminal Court of Appeals Records
   Governor James A. B. Robertson Papers
   Oklahoma State Labor Commission Papers
   State Attorney General's Files
Archives of Labor and Urban Affairs, Walter P. Reuther Library, Wayne State University, Detroit, Michigan
   Alfred and Rose Anderson Collection
   Industrial Workers of the World Collection
   E. W. Latchem Papers, Small Processed Collection
   Pamphlet File
   Fred Thompson Collection
Western History Collections, University of Oklahoma, Norman
   Redmond S. Cole Collection
   Ned DeWitt WPA Writers Project on Oil in Oklahoma
   Indian-Pioneer papers
   Oklahoma State Federation of Labor Collection
   H. C. Peterson Collection

Microfilm Sources

Boehm, Randolph, ed. *United States Military Intelligence Reports: Surveillance of Radicals in the United States, 1917-1941*. Frederick, Md.: University

Publications of America, 1984. This series contains material from National Archives Record Group 165, File Series 10110 and 10058.

Dubofsky, Melvyn, ed. *United States Department of Justice Investigative Files, Part I: The Industrial Workers of the World.* Bethesda, Md.: University Publications of America, 1989. This series contains material from National Archives Record Group 60, Casefiles 186701 and 189152, Bureau of Investigation Files, and from Record Group 204, Casefile 39-242, U.S. Pardon Attorney files.

———. *The Strike Files of the United States Department of Justice, Part I: 1894-1920.* Bethesda, Md.: University Publications of America, 1990.

Naison, Mark D., ed. *United States Department of Justice Investigative Files, Part II: The Communist Party of America.* Bethesda, Md.: University Publications of America, 1989. This series contains material from National Archives Record Group 60, Casefile 202600, Bureau of Investigation Files.

Pardo, Thomas C., ed. *Socialist Collections in the Tamiment Library, New York University, 1872-1956.* New York: Microfilming Corporation of America, 1979.

Socialist Party of America Papers, 1897-1964. Glen Rock, N.J.: Microfilming Corporation of America, 1975.

Southern Tenant Farmers Union Papers, 1934-1970. New York: Microfilming Corporation of America, 1971.

State and Federal Government Publications

Annual Report of the Attorney General of the United States for the Fiscal Year 1918. Washington, D.C.: GPO, 1918.

Annual Report of the Attorney General of the United States for the Fiscal Year 1921. Washington, D.C.: GPO, 1921.

Annual Report of the Attorney General of the United States for the Fiscal Year 1922. Washington, D.C.: GPO, 1922.

Annual Report of the Attorney General of the United States for the Fiscal Year 1923. Washington, D.C.: GPO, 1923.

Annual Report of the Oklahoma State Board of Agriculture 1915. Oklahoma City: Harlow's Printing, 1915.

Arnold, J. H., and R. R. Spafford. "Farm Practices in Growing Wheat." In *Yearbook of the United States Department of Agriculture, 1919*, 123-50. Washington, D.C.: GPO, 1920.

Ashurst, Sen. Harry F. "The IWW Menace." *Congressional Quarterly* 55 (1917): 6687.

BIBLIOGRAPHY

Bizzell, William Bennett. *Farm Tenantry in the United States*. College Station: Texas Agricultural Experiment Station, 1921.

Carney, George. "Slappin' Collars and Stabbin' Pipe: Occupational Folklife of Old-Time Pipeliners." In *Program Book of Smithsonian Festival of American Folklife, 1982*, 15-17. Washington, D.C.: Smithsonian Institution/ GPO, 1982.

Department of Justice. *United States Attorney General's Parole Report*. Washington, D.C.: GPO, 1920.

Eleventh Special Report of the U.S. Commission of Labor. Washington, D.C.: GPO, 1904.

Fowler, H. C. *Accidents in the Petroleum Industry of Oklahoma, 1915-1924*. Department of Commerce Technical Paper no. 392. Washington, D.C.: GPO, 1926.

Fourth Annual Report of the Oklahoma State Department of Labor, 1911. Oklahoma City: Oklahoma Department of Labor, 1911.

Harvey, O. L., ed. *The Anvil and the Plow*. Washington, D.C.: Department of Labor/GPO, 1963.

Horowitz, Daniel. *Labor Relations in the Petroleum Industry*. 2d ed. Works Progress Administration Project no. 465-97-3-7. New York: Works Progress Administration, 1938.

Industrial Relations: Final Report and Testimony Submitted to Congress by the Commission on Industrial Relations Created by the Act of August 23, 1912. 64th Cong., 1st sess., S. Doc. 415, vol. 10. Washington, D.C.: GPO, 1916.

Jamieson, Stuart. *Labor Unionism in American Agriculture*. Department of Labor Bulletin no. 836. Washington, D.C.: GPO, 1945.

Lescohier, Don D. *Harvest Labor Problems in the Wheat Belt*. Department of Agriculture Bulletin no. 1020, April 12, 1922. Washington, D.C.: GPO, 1922.

———. *Conditions Affecting the Demand for Harvest Labor in the Wheat Belt*. Department of Agriculture Bulletin no. 1030, April 1924. Washington, D.C.: GPO, 1924.

Oklahoma State Department of Labor. *Bulletin No. 2*. Oklahoma City: OSDL, 1921.

———. *Oklahoma Department of Labor Bureau of Free Employment Bulletin No. 6*. Oklahoma City: OSDL, 1922.

———. *Bulletin No. 10*. Oklahoma City: OSDL, 1924.

———. *Bulletin No. 10-A*. Oklahoma City: OSDL, 1928.

———. *Annual Report for 1932*. Oklahoma City: OSDL, 1932.

Sixteenth Annual Report of the Commissioner of Labor: Strikes and Lockouts, 1901. Washington, D.C.: GPO, 1901.

Spillman, William J., and E. A. Goldenweiser. "Farm Tenantry in the United States." In *Yearbook of the United States Department of Agriculture, 1916* 321–46. Washington, D.C.: GPO, 1917.

Webb, John N. *The Migratory-Casual Worker*. Works Progress Administration, Division of Social Research, Research Monograph no. 7. Washington, D.C.: GPO, 1937.

Books, Oral Histories, and Memoirs

Ameringer, Oscar. *If You Don't Weaken: The Autobiography of Oscar Ameringer*. New York: Henry Holt and Co., 1940.

Anderson, Nels. *The Hobo: The Sociology of the Homeless Man*. 1923. Reprint, Chicago: Univ. of Chicago Press, 1961.

Ball, Max W. *This Fascinating Oil Business*. Indianapolis: Bobbs-Merrill, 1940.

Bing, Alexander M. *War-Time Strikes and Their Adjustment*. New York: E. P. Dutton, 1921.

Bird, Stewart, Dan Georgakas, and Deborah Shaffer, eds. *Solidarity Forever: An Oral History of the IWW*. Chicago: Lake View Press, 1985.

Boren, Lyle H., and Dale Boren, eds. *Who Is Who in Oklahoma*. Guthrie, Okla.: Co-Operative Publishing Co. 1935.

Brooks, John Graham. *American Syndicalism: The IWW*. New York: Macmillan, 1913.

Chaplin, Ralph. *Wobbly: The Rough-and-Tumble Story of an American Radical*. Chicago: Univ. of Chicago Press, 1949.

Connelly, W. L. *The Oil Business as I Saw It: Half a Century with Sinclair*. Norman: Univ. of Oklahoma Press, 1954.

De Caux, Len. *Labor Radical: From the Wobblies to the CIO, a Personal History*. Boston: Beacon Press, 1970.

Evidence and Cross-Examination of William D. Haywood in the Case of the United States v. William D. Haywood, et al. Chicago: Industrial Workers of the World, 1918.

Foner, Philip S., ed. *Fellow Workers and Friends: IWW Free-Speech Fights as Told by the Participants*. Westport, Conn.: Greenwood Press, 1981.

Foner, Philip S., and Sally M. Miller, eds. *Kate Richards O'Hare: Selected Writings and Speeches*. Baton Rouge: Louisiana State Univ. Press, 1982.

Franks, Kenny, ed. *Voices from the Oil Fields*. Norman: Univ. of Oklahoma Press, 1989.

George, Harris. *The IWW Trial*. Chicago: Industrial Workers of the World, 1919.

Hall, Covington. "Labor Struggles in the Deep South." Unpublished manuscript, photocopy in the Archives of Labor and Urban Affairs, Walter P. Reuther Library, Wayne State University, Detroit.
Harlow, Rex F. *Makers of Government in Oklahoma*. Oklahoma City: Harlow Publishing Company, 1930.
Hill, Luther B. *History of the State of Oklahoma*. Vol. 2. Chicago: Lewis Publishing Company, 1908.
Lampe, William T. *Tulsa County in the World War*. Tulsa: Tulsa County Historical Society, 1919.
Lyons, Eugene. *The Life and Death of Sacco and Vanzetti*. 1927. Reprint, New York: Da Capo, 1970.
———. *Assignment in Utopia*. New York: Harcourt Brace and World, 1938.
Marot, Helen. *American Labor Unions*. New York: Henry Holt, 1914.
McBee, William D. *The Oklahoma Revolution*. Oklahoma City: Modern Publishers, 1956.
Peeler, Lee, ed. *The Standard Blue Book of Oklahoma, 1910-1911*. Oklahoma City: A. J. Peeler and Co., 1909.
Proceedings of the National Convention of the Socialist Party, Chicago, Ill., May 1 to May 6, 1904. Chicago: Socialist Party of America, 1904.
Shannon, C. W. *Handbook on the Natural Resources of Oklahoma*. Norman: Oklahoma Geological Survey, 1916.
Taylor, Paul. *On the Ground in the Thirties*. Layton, Utah: Peregrine Smith, 1983.
Thoburn, Joseph B. *A Standard History of Oklahoma*. Vol. 3. Chicago: American Historical Society, 1916.
Tridon, Andre. *The New Unionism*. New York: B. W. Heubsh, 1913.
Warden-Ebright's Oklahoma City Directory, 1906-1907. Oklahoma City: Warden-Ebright Printing, 1906.

Articles, Chapters, and Pamphlets

Adams, Charles Ely. "The Real Hobo: What He Is and How He Lives." *Forum* 33 (June 1902): 438-49.
"Agricultural Workers, Attention!" *One Big Union Monthly* 1 (April 1919): 53-54.
"Agricultural Workers Industrial Union No. 400 Bulletin No. 29, March 31, 1919." *One Big Union Monthly* 1 (May 1919): 56-57.
Allen, Jack. "A Review of the Facts Relating to the Free-Speech Fight at Minot, North Dakota, from August 1 to August 17, 1913." in *Fellow Workers and Friends: IWW Free-Speech Fights as Told by Participants*, edited by Philip S. Foner, 158-73. Westport, Conn.: Greenwood Press, 1981.

Allen, Henry J. "The New Harvest Hand." *American Review of Reviews* 76 (September 1927): 279–84.

American Civil Liberties Union. "The Truth About the IWW Prisoners." New York: American Civil Liberties Union, 1922. Pamphlet.

Anderson, C. W. "Oil Field Raids." *Defense News Bulletin* 9 (January 1918).

Anderson, Edward E. "Oklahoma Tank Builders Win Two Strikes." *Industrial Solidarity* (December 9, 1922).

Ashleigh, Charles. "The Floater." *International Socialist Review* 15 (July 1914): 34–38.

Barr, Albert. "Oil and Oil Workers." *Industrial Pioneer* 1 (September 1921): 11–12.

"Bills Drafted to Curb the IWW." *Survey* 38 (August 25, 1917): 457–58.

Bohn, William E. "International Notes." *International Socialist Reviews* 17 (April 1917): 629.

Bruere, Robert. "The Industrial Workers of the World: An Interpretation." *Harper's Monthly* 137 (September 1918): 250–57.

Burk, Oran. "From a Cotton Picker." *International Socialist Review* 14 (May 1914): 690.

Carpenter, Ethel E. "The Farmer and His Hired Hand," *Industrial Worker* (Joliet, Ill.) 1 (April 1906): 405.

Carter, T. Lane. "Economic Conditions in the Joplin District." *Engineering and Mining Journal* 90 (October 15, 1910): 759–61.

Chaplin, Ralph. "How the IWW Defends Labor." In *Twenty-Five Years of Industrial Unionism*, 21–28. Chicago: Industrial Workers of the World, 1930.

"Coming Fuel and Transportation Trust," *Industrial Pioneer*, 2d ser., 1 (October 1923): 21

Committee on Public Information, "The Kaiserite in America: One Hundred and One German Lies." Washington, D.C.: GPO, n.d. Forty-page pamphlet.

Connors, Tom. "The Industrial Union in Agriculture." In *Twenty-Five Years of Industrial Unionism*, 35–42. Chicago: Industrial Workers of the World, 1930.

Creel, George, "Harvesting the Harvest Hands." *Harper's Weekly*, September 26, 1914, 292–94.

———. "The American Newspaper." *Everybody's Magazine* 40 (April 1919): 40–44, 92.

Dimmit, William, "An Organized Harvest," *Industrial Pioneer* 1 (September 1921): 3–5.

———. "Splitting the Big Drive," *Industrial Pioneer* 1 (December 1921): 11.

Doree, E. F. "Gathering the Grain." *International Socialist Review* 15 (June 1918): 740-43.

Embree, A. S. "Introduction." In Harrison George, *The IWW Trial*, 3-13. Chicago: Industrial Workers of the World, 1919.

"Emptying the School to Work the Farm," *Survey* 38 (September 29, 1917): 576.

Feinberg, Harry. "Defense News." *Industrial Pioneer* 1 (August 1921): 60.

Flynn, Elizabeth Gurley, "The Free-Speech Fight at Spokane." In *Fellow Workers and Friends: IWW Free-Speech Fights as Told by the Participants*, edited by Philip S. Foner, 34-55. Westport, Conn.: Greenwood Press, 1981.

Foote, E. J. "Break Ranks! Come with Us!" *Industrial Worker* (Joliet, Ill.) 1 (June 1906): 3.

———. "A Voice from the Ranks," *Industrial Worker* (Joliet, Ill.) 1 (August 1906): 1-2.

Forbes, James. "The Tramp, or Caste in the Jungle." *Outlook* 98 (August 1911): 869-75.

"Forty Years of the Oil Age." New York: American Petroleum Institute, 1944. Booklet.

Fox, Mat K. "Agricultural Workers Industrial Union No. 400 Bulletin No. 29, March 31, 1919." *One Big Union Monthly* (May 1919): 56.

———. "The Story of No. 400." *One Big Union Monthly* 1 (September 1919): 49.

Fox, Mat K., and Walter Sheridan. "Agricultural Workers Industrial Union No. 400, IWW." *One Big Union Monthly* 1 (March 1919): 58-59.

Foy, James. "A Migratory Worker's Diary." *Industrial Pioneer* 1 (February 1924): 29.

Friedman, I. K. "Our Inland Migrations." *World's Work* 8 (September 1904): 5287-90.

Hader, John J. "Honk Honk Hobo." *Survey* 60 (August 1, 1928): 453-55.

Hanson, Nils H. "Among the Harvesters." *International Socialist Review* 16 (August 1915): 75-78.

Holbrook, Stuart H. "The Last of the Wobblies." In *Little Annie Oakley and Other Rugged People*, by Stuart H. Holbrook, 167-76. New York: Macmillan, 1948.

Holman, Charles W. "The Tenant Farmer, Country Brother of the Casual Worker." *Survey* 34 (April 17, 1915): 62-64.

Hoxie, Robert F. "The Industrial Workers of the World and Revolutionary Unionism." In *Readings in Labor Economics and Labor Relations*, rev. ed., edited by Richard L. Rowan, 129-43. Homewood, Ill.: Richard L. Irwin, 1972.

"In the Domain of Standard Oil." *Industrial Pioneer*, 2d ser., 1 (September 1923): 17-19

"In the Oil Fields." *International Socialist Review* 14 (May 1914): 664-67.

"International Notes." *International Socialist Review* 17 (April 1914): 629.

"John J. Carney." in *History of the State of Oklahoma*, vol. 2, edited by Luther B. Hill. Chicago: Lewis Publishing Company, 1908.

Lane, Winthrop D. "Uncle Sam, Jailer." *Survey* 42 (September 6, 1919): 800-12, 834.

Latchem, E. W. "The Modern Agricultural Slave." *One Big Union Monthly* 2 (August 1920): 54-57.

———. "The Agricultural Workers Convention" and "Aftermath." *One Big Union Monthly* 2 (November 1920): 56-58.

Lescohier, Don D. "Hands and Tools of the Wheat Harvest." *Survey* 50 (July 1, 1923): 376-82, 409-10, 412.

———. "Harvesters and Hoboes in the Wheat Fields." *Survey* 50 (August 1, 1923): 482-87, 503-504.

Lewis, O. F. "The Tramp Problem." *Annals of the American Academy of Political and Social Sciences* 40 (March 1912): 217-27.

Lowe, Caroline A. "A Letter from Our Attorney on the Wichita Case." *One Big Union Monthly* 1 (September 1919): 9.

Lyons, Eugene. "Tulsa: A Study in Oil." *One Big Union Monthly* 1 (December 1919): 35-37.

Mackey, C. J. "The Industrial Workers of the World vs. Any Other Kind of Unionism." *Industrial Worker* 1 (February 1906): 1

Marcosson, Isaac F. "Harvesting the Wheat." *World's Work* 9 (November 1904): 54-59.

McCook, John J. "A Tramp Census and Its Revelations." *Forum* 15 (August 1893): 753-56.

Mills, Charles Morris. "Joplin Zinc: Industrial Conditions in the World's Greatest Zinc Center." *Survey* 45 (February 5, 1921): 657-66.

"Mob Violence in the United States." *Survey* 40 (April 27, 1918): 101-102.

National Civil Liberties Bureau. "The 'Knights of Liberty' Mob and the IWW Prisoners at Tulsa, Oklahoma, November 9, 1917." New York: National Civil Liberties Bureau, February 1918. Pamphlet.

News and Views. *International Socialist Review* 17 (January 1917): 443.

———. *International Socialist Review* 17 (March 1917): 573.

———. *International Socialist Review* 18 (December 1917): 315-16.

———. *International Socialist Review* 18 (January 1918): 378.

Nolan, Ed, "From Frisco to Denver." in *Fellow Workers and Friends: IWW Free-Speech Fights as Told by Participants,* edited by Philip S. Foner, 146-52. Westport, Conn.: Greenwood Press, 1981.

"Organizing the Oil Industry No. 5—the Tank Builders." *Industrial Solidarity* (August 6, 1921).

Ostrander, Joseph. "The New Labor Displacers." *Industrial Pioneer,* 2d ser., 2 (July 1924): 31-32, 34.

Pannell, W. W. "Tenant Farming in the United States." *International Socialist Review* 16 (January 1916): 421-23.

Perry, Grover H. "The Revolutionary IWW." Cleveland: Industrial Workers of the World Publications Bureau, July 1916. Pamphlet.

Simpson, D. N., and Mat K. Fox. "Agricultural Workers Industrial Union No. 110 Bulletin No. 32, May 5, 1919." *One Big Union Monthly* 1 (June 1919): 56-57.

———. "Agricultural Workers Industrial Union No. 110 Bulletin No. 37, June 9, 1919." *One Big Union Monthly* (July 1919): 57.

———. "Agricultural Workers Industrial Union No. 110 Bulletin No. 45, August 11, 1919." *One Big Union Monthly* 1 (September 1919): 50.

———. "Agricultural Workers Industrial Union No. 110 Bulletin No. 56, November 10, 1919." *One Big Union Monthly* 1 (December 1919): 53-54.

Smith, Walker C. "Sabotage: Its History, Philosophy, and Function." Spokane: Smith, 1913. Pamphlet.

"The Anti-Syndicalist Laws." *One Big Union Monthly* 1 (April 1919): 9.

"The Campaign of the Agricultural Workers." *One Bit Union Monthly* 1 (August 1919): 7-8.

"The General Defense Committee of the IWW." *International Socialist Review* 18 (February 1918): 408-409.

Veblen, Thorstein, "On the Nature and Uses of Sabotage." *Dial* 66 (April 5, 1919): 341-46.

Wells, Nick, "As to Lubrication." *Industrial Pioneer* 1 (November 1921): 22-23.

———. "Tactics in Oil." *Industrial Pioneer* 1 (December 1921): 26.

"What Do You Think of This?" Chicago: Industrial Workers of the World, 1918. Four-page pamphlet.

Workman, E. "History of the 400 A.W.O.: The One Big Union Idea in Action." New York: One Big Union Club, 1939. Pamphlet.

Wright, Hamilton, "Harvesting the West's Great War Crops." *Forum* 58 (October 1917): 481-90.

Newspapers

IWW Publications

Industrial Pioneer (Chicago)
Industrial Solidarity (Chicago)
Industrial Union Bulletin (Chicago)
Industrial Worker (Joliet, Ill., 1906-1907)
Industrial Worker (Spokane and Seattle, Wash., 1909-17)
New Solidarity (Chicago)
One Big Union Monthly (Chicago)
Rebel Worker (New York)
Solidarity (New Castle, Penn.; Cleveland, Ohio; Chicago)

Labor and Radical Publications

American Labor Union Journal (Butte, Mont.)
American Socialist (Chicago)
Beckham County Advocate (Carter)
Grant County Socialist (Medford)
Industrial Democrat (Oklahoma City)
Industrial Unionist (New York)
Industrial Union News (Detroit)
International Socialist Review (Chicago)
Journal of United Labor (Baltimore)
Labor Amalgamator (Muskogee)
Labor World (Shawnee)
Medford Challenge
Miners' Magazine (Denver)
New Century (Sulphur)
Oklahoma Federationist (Oklahoma City)
Oklahoma Labor (Oklahoma City)
Oklahoma Labor Unit (Oklahoma City)
Oklahoma Leader (Oklahoma City)
Oklahoma Pioneer (Oklahoma City)
Otter Valley Socialist (Snyder)
Party Builder (Chicago)
Social Democrat (Oklahoma City)
Tulsa Labor News
Unionist (Muskogee)
Voice of Labor (Chicago—American Labor Union)

Other Newspapers and Magazines

Ada Weekly News
Cushing Citizen
Daily Oklahoma State Capital (Guthrie)
Daily Oklahoma Times-Journal (Oklahoma City)
Daily Oklahoman (Oklahoma City)
Drumright Daily News (Oklahoma)
Drumright Derrick and Drumright Evening Derrick (Oklahoma)
Drumright Evening News
El Reno News
Enid Daily Eagle
Enid Daily News
Enid Morning News
Eufala Indian Journal
Guthrie Daily Leader
Harlow's Weekly (Oklahoma City)
Holdenville Times Democrat
Imperial Nighthawk (Atlanta)
Indian Journal (Muskogee, I. T.)
Kansas City Times
Lexington Leader
McAlester News-Capital
Miami (Oklahoma) Daily Record-Herald
New York Times
Oil and Gas Journal (Tulsa)
Oklahoma City Times
Oklahoma Farmer-Stockman (Oklahoma City)
Oklahoma News (Oklahoma City)
Oklahoma War Chief (Caldwell, Kans.)
Pittsburgh County Guardian
Seminole County News
Shawnee News Herald
Stillwater Advance
Stillwater Gazette
Tulsa Daily World
Tulsa Democrat
Tulsa Times
Tulsa Tribune

Tulsa World
Vinita Weekly Chieftain
Wewoka Democrat

Court Cases

Trial record and transcript, *State of Oklahoma v. Arthur Berg*, Criminal Case No. 2146, County District Court, McAlester

Berg v. State, 29 Oklahoma Criminal Court of Appeals 112, 233 Pacific Reporter 497

Gilbert v. Minnesota, 254 U.S. 325

Gitlow v. New York, 268 U.S. 562

Transcript of record, United States Court of Appeals, Eighth Circuit, no. 5170, Clure Isenhour et al., plaintiff in error, v. United States, defendant in error, June 1, 1918

Jaffee v. State 76 Oklahoma Criminal Court of Appeals 95

State of Oklahoma v. Charles Krieger, Criminal Case No. 1576, County District Court, Tulsa

Shaw v. State, 76 Oklahoma Criminal Court of Appeals 305

Transcript of record, *State of Oklahoma v. Grover Jackson Terrell*, Criminal Case No. 520, County District Court, Enid

Terrell v. State, 217 Padivid Reporter 900

Homer Wear v. State of Oklahoma, transcript and brief of plaintiff in error, March 25, 1925, County District Court, Miami; 30 Oklahoma Criminal Court of Appeals 118, 235 Pacific Reporter 271

Ina Wood v. State, 76 Oklahoma Criminal Court of Appeals 89

Robert Wood v. State, 77 Oklahoma Criminal Court of Appeals 305

SECONDARY SOURCES

Dissertations, Theses, and Unpublished Material

Aldrich, Gene. "A History of the Coal Industry in Oklahoma to 1907." Ph.D. dissertation, University of Oklahoma, 1952.

Alexander, Charles Comer. "Invisible Empire in the Southwest: The Ku _lux Klan in Texas, Louisiana, Oklahoma, and Arkansas in 1920–1930." Ph.D. dissertation, University of Texas, 1962.

Applen, Allen Gale. "Migratory Harvest Labor in the Midwestern Wheat Belt, 1870–1940." Ph.D. dissertation, Kansas State University, 1974.

Asfahl, Milton Ernest. "Oklahoma and Organized Labor." Master's thesis, University of Oklahoma, 1930.

Buhle, Paul M. "Marxism in the United States." Ph.D. dissertation, University of Wisconsin-Madison, 1975.

Bush, Charles. "The Green Corn Rebellion." Master's thesis, University of Oklahoma, 1932.
Deberry, Drue Lemeul. "The Ethos of the Oklahoma Oil Boom Frontier." Master's thesis, University of Oklahoma, 1970.
Fast, Stanley P. "The Agricultural Workers Organization and the Harvest Stiff in the Midwestern Wheat Belt, 1915-1920." Master's thesis, Mankato State University, 1974.
Gilbert, James Leslie. "Three Sands: Oklahoma Oil Field and Community of the 1920s." Master's thesis, University of Oklahoma, 1967.
Graham, Donald Ralph. "Red, White, and Black: An Interpretation of Ethnic and Racial Attitudes of Agrarian Radicals in Texas and Oklahoma, 1880-1920." Master's thesis, University of Saskatchewan-Regina, 1973.
Harder, Jesse. "Wheat Production in Northwestern Oklahoma, 1893-1932." Master's thesis, University of Oklahoma, 1952.
Parker, Albert Raymond. "Life and Labor in the Mid-Continent Oil Fields." Ph.D. dissertation, University of Oklahoma, 1951.
Pope, Virginia C. "The Green Corn Rebellion: A Case Study of Newspaper Self-Censorship." Master's thesis, Oklahoma A&M College, 1940.
Reinhart, Sister M. Magdalen, O.S.B. "The Open Shop Movement in Oklahoma." Master's thesis, University of Oklahoma, 1938.
Rosen, Ellen I. "Peasant Socialism in America? The Socialist Party in Oklahoma before the First World War." Ph.D. dissertation, City University of New York, 1975.
Sellars, Nigel Anthony. "Butchers and Businessmen: The 1921 Packinghouse Strike in Oklahoma City." Paper presented at the annual meeting of the Southwest Social Science Association, Fort Worth, Texas, March 22, 1990.
———. "Miners, Farmers, and Politicians: The Knights of Labor in Oklahoma, 1882-1894." Paper presented at the Oklahoma regional conference of Phi Alpha Theta, University of Oklahoma, Norman, March 27, 1993.
Smith, James Morton. "Criminal Syndicalism in Oklahoma: A History of the Law and Its Application." Master's thesis, University of Oklahoma, 1946.
Snodgrass, William George. "A History of the Oklahoma State Federation of Labor to 1918." Master's thesis, University of Oklahoma, 1960.
Steers, Nina A. "The Ku Klux Klan in Oklahoma in the 1920s." Master's thesis, Columbia University, 1965.

Warrick, Sherry Harrod. "Antiwar Reaction in the Southwest during World War I." Master's thesis, University of Oklahoma, 1973.
White, Earl Bruce. "The Wichita Indictments and Trial of the Industrial Workers of the World, 1917-1919, and the Aftermath." Ph.D. dissertation, University of Colorado, 1980.
Womack, John, Jr. "Oklahoma's Green Corn Rebellion: The Importance of Fools." Senior thesis, Yale University, 1959.

Books

Allen, Ruth A. *Chapters in the History of Organized Labor in Texas*. Publication no. 4143. Austin: University of Texas Bureau of Research in the Social Sciences, 1941.
Brissenden, Paul F. *The IWW: A Study of American Syndicalism*. 1920. Reprint, New York: Russell and Russell, 1957.
Burbank, Garin. *When Farmers Voted Red: The Gospel of Socialism in the Oklahoma Countryside, 1910-1924*. Westport, Conn.: Greenwood Press, 1976.
Chalmers, David M. *Hooded Americanism: The History of the Ku Klux Klan*. New York: New Viewpoints, 1981.
Clark, J. Stanley. *The Oil Century*. Norman: Univ. of Oklahoma Press, 1958.
Collings, Ellsworth, and Alma Miller England. *The 101 Ranch*. 1937. Reprint, Norman: Univ. of Oklahoma Press, 1971.
Conlin, Joseph. *Bread and Roses, Too: Studies of the Wobblies*. Westport, Conn.: Greenwood Press, 1969.
Davis, Mike. *Prisoners of the American Dream: Politics and Economy in the History of the U.S. Working Class*. London: Verso, 1986.
Debo, Angie. *The Rise and Fall of the Choctaw Republic*. 2d ed. Norman: Univ. of Oklahoma Press, 1961.
———, and John M. Oskison, eds. *Oklahoma: A Guide to the Sooner State*. Norman: Univ. of Oklahoma Press, 1941.
Dick, William M. *Labor and Socialism in America: The Gompers Era*. Port Washington, N.Y.: Kennikat Press, 1972.
Dowell, Eldridge Foster. *A History of Criminal Syndicalism Legislation in the United States*. Johns Hopkins University Studies in Historical and Political Science, ser. 56, no. 1. Baltimore: Johns Hopkins Univ. Press, 1939.
Drache, Hiram. *Day of the Bonanza*. Fargo: North Dakota Institute of Regional Studies, 1964.
Dubofsky, Melvyn. *We Shall Be All: A History of the Industrial Workers of the World*. 2d ed. Urbana: Univ. of Illinois Press, 1988.

Ellsworth, Scott. *Death in a Promised Land: The Tulsa Race Riot of 1921*. Baton Rouge: Louisiana State Univ. Press, 1982.
Federal Writers Project of Oklahoma, Works Progress Administration. *Labor History of Oklahoma*. Oklahoma City: A. M. Van Horn, 1939.
Filene, Peter. *Him/Her/Self: Sex Roles in Modern Society*. Rev. ed. Baltimore: Johns Hopkins Univ. Press, 1992.
Fite, Gilbert C. *The Farmers' Frontier, 1865-1900*. 1966. Reprint, Norman: Univ. of Oklahoma Press, 1987.
Foner, Philip S. *History of the Labor Movement in the United States*. Vol. 2, *From the Founding of the American Federation of Labor to the Emergence of American Imperialism*. New York: International Publishers, 1955.
———. *History of the Labor Movement in the United States*. Vol. 3, *The Policies and Practices of the American Federation of Labor, 1900-1909*. New York: International Publishers, 1964.
———. *History of the Labor Movement in the United States*. Vol. 4, *The Industrial Workers of the World, 1905-1917*. New York: International Publishers, 1965.
Forbath, William. *Law and the Shaping of the American Labor Movement*. Cambridge: Harvard Univ. Press, 1991.
Franklin, Jimmie Lewis. *Born Sober: Prohibition in Oklahoma, 1907-1959*. Norman: Univ. of Oklahoma Press, 1971.
Galenson, Walter. *The CIO Challenge to the AFL: A History of the Labor Movement, 1935-1941*. Cambridge: Harvard Univ. Press, 1960.
Gambs, John S. *The Decline of the IWW*. 1932. Reprint, New York: Russell and Russell, 1966.
Garlock, Jonathan. *A Guide to the Local Assemblies of the Knights of Labor*. Westport, Conn.: Greenwood Press, 1982.
Gibson, Arrell Morgan. *Oklahoma: A History of Five Centuries*. 2d ed. Norman: Univ. of Oklahoma Press, 1981.
———. *Wilderness Bonanza: The Tri-State District of Missouri, Kansas, and Oklahoma*. Norman: Univ. of Oklahoma Press, 1972.
Glasscock, C. B. *Then Came Oil: The Story of the Last Frontier*. 1938. Reprint, Westport, Conn.: Greenwood Press, 1976.
Green, Archie. *Wobblies, Pile Butts, and Other Heroes: Laborlore Explorations*. Urbana: Univ. of Illinois Press, 1993.
Green, James R. *Grass-Roots Socialism: Radical Movements in the Southwest, 1895-1943*. Baton Rouge: Louisiana State Univ. Press, 1978.
———. *The World of the Worker: Labor in Twentieth-Century America*. New York: Hill and Wang, 1980.
Handlin, Oscar, and Mary Handlin. *The Dimensions of Liberty*. Cambridge: Harvard Univ. Press, Belknap Press, 1961.

Hattam, Victoria C. *Labor Visions and State Power: The Origins of Business Unionism in the United States.* Princeton, N.J.: Princeton Univ. Press, 1993.

Hicks, John D. *The Populist Revolt: A History of the Farmers' Alliance and the People's Party.* 1931. Reprint, Lincoln: Univ. of Nebraska Press, 1961.

Higham, John. *Strangers in the Land: Patterns of American Nativism, 1860–1925.* New York: Atheneum, 1969.

Hobsbawm, Eric J. *Primitive Rebels: Studies in Archaic Forms of Social Movements in the Nineteenth and Twentieth Centuries.* New York: W. W. Norton, 1959.

———. *Workers: Worlds of Labor.* New York: Pantheon, 1984.

Isern, Thomas D. *Custom Combining on the Great Plains: A History.* Norman: Univ. of Oklahoma Press, 1981.

———. *Bull Threshers and Bindlestiffs: Harvesting and Threshing on the North American Plains.* Lawrence: Univ. Press of Kansas, 1990.

Jackson, Kenneth T. *The Ku Klux Klan in the City, 1915–1930.* New York: Oxford Univ. Press, 1967.

Jacoby, Sanford M. *Employing Bureaucracy: Managers, Unions, and the Transformation of Work in American Industry, 1900–1945.* New York: Columbia Univ. Press, 1985.

Jensen, Joan M. *The Price of Vigilance.* New York: Rand McNally, 1968.

———. *Army Surveillance in America, 1775–1980.* New Haven: Yale Univ. Press, 1991.

Jenson, Vernon H. *Heritage of Conflict: Labor Relations in Nonferrous Metals Industry.* Ithaca, N.Y.: Cornell Univ. Press, 1950.

Johnson, Charles S., Edwin R. Embree, and William Alexander. *The Collapse of Cotton Tenancy.* Chapel Hill: Univ. of North Carolina Press, 1935.

Karson, Mark. *American Labor Unions and Politics, 1900–1918.* Carbondale: Southern Illinois Univ. Press, 1958.

Kennedy, David M. *Over Here: The First World War and American Society.* New York: Oxford Univ. Press, 1980.

Kipnis, Ira. *The American Socialist Movement, 1897–1912.* New York: Columbia Univ. Press, 1952.

Kornbluh, Joyce, ed. *Rebel Voices: An IWW Anthology.* Ann Arbor: Univ. of Michigan Press, 1964.

Lindsey, Almont. *The Pullman Strike.* Chicago: Univ. of Chicago Press, 1942.

Lipset, Seymour Martin, Martin Trow, and James Coleman. *Union Democracy.* New York: Anchor Books, 1962.

Mathewson, Stanley B. *Restriction of Output among Unorganized Workers.* New York: Viking Press, 1931.

McConnell, Grant. *The Decline of Agrarian Democracy*. Berkeley and Los Angeles: Univ. of California Press, 1959.

McLean, John G., and Robert William Haigh. *The Growth of Integrated Oil Companies*. Boston: Harvard University Graduate School of Business Administration, 1954.

McWilliams, Carey. *Ill Fares the Land: Migrants and Migratory Labor in the United States*. Boston: Little, Brown, 1942.

Miller, Worth Robert. *Oklahoma Populism: A History of the People's Party in the Oklahoma Territory*. Norman: Univ. of Oklahoma Press, 1987.

Mock, James R., and Evangeline Thurber. *Report on Demobilization*. Norman: University of Oklahoma, 1944.

Montgomery, David. *Workers' Control in America: Studies in the History of Work, Technology, and Labor Struggles*. New York: Cambridge Univ. Press, 1979.

Morris, John W. *Ghost Towns of Oklahoma*. Norman: Univ. of Oklahoma Press, 1978.

Murphy, Paul L. *World War I and the Origins of Civil Liberties in the United States*. New York: W. W. Norton, 1979.

Newsom, D. Earl. *Drumright! The Glory Days of a Boom Town*. Perkins, Okla.: Evans Publications, 1985.

———. *Drumright II: A Thousand Memories*. Perkins, Okla.: Evans Publications, 1985.

Noggle, Burl. *Into the Twenties: The United States from Armistice to Normalcy*. Urbana: Univ. of Illinois Press, 1974.

O'Connor, Harvey. *History of the Oil Workers International Union-CIO*. Denver: OWIU-CIO, 1950.

Olien, Roger, and Diana Davids Olien. *Oil Boom: Social Change in Five Texas Towns*. Lincoln: Univ. of Nebraska Press, 1982.

Pollack, Norman. *The Populist Response to Industrial America: Midwestern Political Thought*. Cambridge: Harvard Univ. Press, 1962.

Preston, William, Jr. *Aliens and Dissenters: Federal Suppression of Radicals, 1903-1933*. Cambridge: Harvard Univ. Press, 1963.

Renshaw, Patrick. *The Wobblies: The Story of Syndicalism in the United States*. Garden City, N.Y.: Doubleday, 1967.

Ringenbach, Paul T. *Tramps and Reformers, 1873-1916: The Discovery of Unemployment in New York*. Westport, Conn.: Greenwood Press, 1973.

Rister, Carl Coke. *Oil! Titan of the Southwest*. Norman: Univ. of Oklahoma Press, 1949.

Robinson, James A. *Anti-Sedition Legislation and Loyalty Investigations in Oklahoma*. Norman: University of Oklahoma Bureau of Government Research, 1956.

Rodgers, Daniel T. *The Work Ethic in Industrial America, 1850–1920*, Chicago: Univ. of Chicago Press, 1978.

Ryan, Frederick Lynne. *The Rehabilitation of Oklahoma Coal Mining Communities*. Norman: Univ. of Oklahoma Press, 1935.

———. *Problems of the Oklahoma Labor Market, with Special Reference to Unemployment Compensation*. Oklahoma City: Semco Color Press, 1937.

Saposs, David J. *Left-Wing Unionism: A Study of Radical Policies and Tactics*. New York: International Publishers, 1926.

Sautter, Udo. *Three Cheers for the Unemployed: Government and Unemployment before the New Deal*. New York: Cambridge Univ. Press, 1991.

Scales, James R., and Danney Goble. *Oklahoma Politics: A History*. Norman: Univ. of Oklahoma Press, 1982.

Schob, David E. *Hired Hands and Plowboys: Farm Labor in the Midwest, 1815–1860*. Urbana: Univ. of Illinois Press, 1975.

Sexton, Patricia Cayo. *The War on Labor and the Left: Understanding America's Unique Conservatism*. Boulder, Colo.: Westview Press, 1991.

Shannon, David A. *The Socialist Party of America: A History*. New York: Macmillan, 1955.

Shirk, George. *Oklahoma Place-Names*, 2d ed. Norman: Univ. of Oklahoma Press, 1974.

Sinclair, Upton. *The Brass Check: A Study of American Journalism*. Rev. ed. Pasadena, Calif.: Sinclair, 1931.

Solberg, Carl. *Oil Power: The Rise and Imminent Fall of an American Empire*. New York: Mentor Books, 1976.

Stearns, Peter N. *Be a Man! Males in Modern Society*. New York: Holmes and Meier Publishers, 1979.

Suggs, George G., Jr. *Union Busting in the Tri-State: The Oklahoma, Kansas, and Missouri Metal Workers Strike of 1935*. Norman: Univ. of Oklahoma Press, 1986.

Talbert, Roy, Jr. *Negative Intelligence: The Army and the American Left, 1917–1941*. Jackson: Univ. Press of Mississippi, 1991.

Tatum, Georgia Lee. *Disloyalty in the Confederacy*. Chapel Hill: Univ. of North Carolina Press, 1934.

Tax, Meredith. *The Rising of the Women: Feminist Solidarity and Class Conflict, 1880–1917*. New York: Monthly Review Press, 1980.

Thompson, E. P. *The Making of the English Working Class*. New York: Vintage, 1966.

———. *Whigs and Hunters: The Origins of the Black Act*. New York: Pantheon, 1975.

Thompson, Fred, and Patrick Murfin. *The IWW: Its First Seventy Years*. Chicago: Industrial Workers of the World, 1976.
Thompson, John. *Closing the Frontier: Radical Response in Oklahoma, 1889-1923*. Norman: Univ. of Oklahoma Press, 1986.
Tindall, George Brown. *The Emergence of the New South, 1913-1945*. Baton Rouge: Louisiana State Univ. Press, 1967.
Trachtenberg, Alan. *The Incorporation of America: Culture and Society in the Gilded Age*. New York: Hill and Wang, 1982.
Tyler, Robert L. *Rebels of the Woods: The IWW in the Pacific Northwest*. Eugene: Univ. of Oregon Press, 1967.
Voss, Kim. *The Making of American Exceptionalism: The Knights of Labor and Class Formation in the Nineteenth Century*. Ithaca, N.Y.: Cornell Univ. Press, 1993.
Weinstein, James. *The Decline of Socialism in America, 1912-1925*. 1967. Reprint, New Brunswick, N.J.: Rutgers Univ. Press, 1984.
White, Richard. *"It's Your Misfortune and None of My Own": A New History of the American West*. Norman: Univ. of Oklahoma Press, 1991.
Wiebe, Robert H. *Businessmen and Reform: A Study of the Progressive Movement*. Cambridge: Harvard Univ. Press, 1962.
―――. *The Search for Order, 1877-1920*. New York: Hill and Wang, 1967.
Wilentz, Sean. *Chants Democratic: New York City and the Rise of the American Working Class, 1788-1850*. New York: Oxford Univ. Press, 1984.
Winters, Donald E., Jr. *The Soul of the Wobblies: The IWW, Religion, and American Culture in the Progressive Era, 1905-1917*. Westport, Conn.: Greenwood Press, 1985.
Womack, John, Sr. *Annals of Cleveland County, Oklahoma, 1889-1957*. Norman: Transcript Press, 1976.
―――. *Norman: An Early History, 1820-1900*. Norman: Womack, 1976.
Worster, Donald. *The Dust Bowl: The Southern Plains in the 1930s*. New York: Oxford Univ. Press, 1979.
Wyatt-Brown, Bertram. *Honor and Violence in the Old South*. New York: Oxford Univ. Press, 1986.
Zieger, Robert H. *American Workers, American Unions, 1920-1985*. Baltimore: Johns Hopkins Univ. Press, 1986.

Articles and Chapters

Agnew, Brad, "Wagoner, I.T.: Queen City of the Prairies." *Chronicles of Oklahoma* 64 (winter 1986-87): 16-47.
Barnhill, John. "Triumph of Will: The Coal Strike of 1899-1903." *The Chronicles of Oklahoma* 61 (spring 1983): 80-95.

Brissenden, Paul F. "The Industrial Workers of the World." In *Encyclopedia of the Social Sciences*, vol. 8, edited by Edwin R. A. Seligman, 13-18. New York: Macmillan, 1930.

Brody, David. "Labor Movement." In *Encyclopedia of American Political History: Studies of Principal Movements and Ideas*, vol. 2, edited by Jack Greene, 709-27. New York: Scribners, 1984.

Brown, Richard Maxwell. "The History of Extralegal Violence in Support of Community Values." In *Violence in America: A Historical and Contemporary Reader*, edited by Thomas Rose, 86-95. New York: Vintage Books, 1970.

———. "Historical Patterns of American Violence." In *Violence in America: Historical and Comparative Perspectives*, rev. ed., edited by Hugh Davis Graham and Ted Robert Gurr, 19-48. Beverly Hills, Calif.: Sage Publications, 1979.

———. "The American Vigilante Tradition." In *Violence in America: Historical and Comparative Perspectives*, rev. ed., edited by Hugh Davis Graham and Ted Robert Gurr, 153-85. Beverly Hills, Calif.: Sage Publications, 1979.

———. "Law and Order on the American Frontier: The Western Civil War of Incorporation." In *Law for the Elephant, Law for the Beaver: Essays in the Legal History of the North American West*, edited by John McLaren, Hamar Foster, and Chet Orloff, 74-89. Regina, Saskatchewan: Canadian Plains Research Center/University of Regina, 1992.

———. "Western Violence: Structures, Values, Myth." *Western Historical Quarterly* 24 (February 1993): 5-20.

Bryant, Keith L., "Labor in Politics: The Oklahoma State Federation of Labor during the Age of Reform." *Labor History* 11 (summer 1970): 259-76.

Burbank, Garin. "Socialism in an Oklahoma Boom-Town: 'Milwaukeeizing' Oklahoma City." In *Socialism and the Cities*, edited by Bruce M. Stave, 99-115. Port Washington, N.Y.: Kennikat Press, 1975.

Carney, George. "The Historic Preservation of the Cushing Oil Field: A Summary." *Payne County Historical Review* 1 (April 1981): 3-11.

Clark, J. Stanley. "Texas Fever in Oklahoma." *Chronicles of Oklahoma* 24 (winter 1951/52): 429-43.

Crockett, Norman L. "The Opening of Oklahoma: A Businessman's Frontier." *Chronicles of Oklahoma* 55 (spring 1978): 85-95.

Dancis, Bruce. "Socialism and Women in the United States, 1900-1917." *Socialist Revolution* 6 (January-March 1976): 81-144.

Davis, Mike. "Forced to Tramp." In *Walking to Work: Tramps in America, 1790-1935*, edited by Eric Monkkonen, 142-70. Lincoln: Univ. of Nebraska Press, 1984.

Doherty, Robert E. "Thomas J. Hagerty, the Church, and Socialism." *Labor History* 3 (winter 1962): 39-56.

Dubofsky, Melvyn. "The Origins of Western Working Class Radicalism, 1890-1905." In *The Labor History Reader*, edited by Daniel J. Leab, 230-53. Urbana: Univ. of Illinois Press, 1985.

Fink, Leon. "Labor, Liberty, and the Law: Trade Unionism and the Problem of the American Constitutional Order." *Journal of American History* 74 (December 1987): 904-25.

Fowler, James H., Jr. "Tar and Feather Patriotism: The Suppression of Dissent in Oklahoma During World War I." *Chronicles of Oklahoma* 56 (winter 1978-79): 409-30.

———. "Creating an Atmosphere of Suppression, 1914-1917." *Chronicles of Oklahoma* 59 (summer 1981): 202-23.

Gibson, Arrell Morgan. "Oklahoma: Land of the Drifter, Deterrents to Sense of Place." *Chronicles of Oklahoma* 64 (summer 1986): 5-13.

Green, Donald E. "Beginnings of Wheat Culture in Oklahoma." In *Rural Oklahoma*, edited by Donald E. Green, 56-73, Oklahoma City: Oklahoma Historical Society, 1977.

Green, James R. "The Brotherhood." In *Working Lives: the Southern Exposure History of Labor in the South*, edited by Marc S. Millers, 22-39. New York: Pantheon, 1980.

Gutman, Herbert G. "The Labor Policies of the Large Corporation in the Gilded Age: The Case of the Standard Oil Company." In Herbert G. Gutman, *Power and Culture: Essays on the American Working Class*, edited by Ira Berlin, 213-54. New York: Pantheon Books, 1987.

———. "The Workers' Search for Power: Labor in the Gilded Age." In Herbert G. Gutman, *Power and Culture: Essays on the American Working Class*, edited by Ira Berlin, 70-92. New York: Pantheon Books, 1987.

Hale, Douglas. "The People of Oklahoma: Economic and Social Change." In *Oklahoma: New Views of the Forty-Sixth State*, edited by Anne Hodges Morgan and H. Wayne Morgan, 31-92. Norman: Univ. of Oklahoma Press, 1982.

Hilton, O. A. "The Oklahoma Council of Defense and the First World War." *Chronicles of Oklahoma* 20 (March 1942): 18-42.

Hobby, Daniel T. " 'We Have Got Results': A Document on the Organization of Domestics in the Progressive Era." *Labor History* 17 (winter 1976): 104-107.

Jameson, Elidabeth. "Imperfect Unions: Class and Gender in Cripple Creek, 1894-1904." In *Class, Sex, and the Woman Worker*, edited by Milton Cantor and Bruce Laurie, 166-202. Westport, Conn.: Greenwood Press, 1977.

Karsten, Peter. "Armed Progressives: The Military Reorganizes for the American Century." In *The Military in America: From the Colonial to the Present*, edited by Peter Karsten, 229-71. New York: Free Press, 1980.

Kerr, Clark, and Abraham Siegel. "The Interindustry Propensity to Strike: An International Comparison." In *Industrial Conflict*, edited by Arthur Kornhauser, Robert Dubin, and Arthur M. Ross, 189-212. New York: McGraw-Hill, 1954.

Kingsdale, Jon M. "'The Poor Man's Club': Social Functions of the Urban Working Class Saloon." In *The American Male*, edited by Elizabeth H. Pleck and Joseph H. Pleck, 255-83. Englewood Cliffs, N.J.: Prentice-Hall, 1980.

Koppes, Clayton R. "The Kansas Trial of the IWW, 1917-1919." *Labor History* 16 (summer 1975): 339-58.

McGinnis, Patrick. "'Share the Work': Ira M. Finley and the Veterans of Industry of America." In *Hard Times in Oklahoma: The Depression Years*, edited by Kenneth D. Hendrickson, Jr., 22-46. Oklahoma City: Oklahoma Historical Society, 1983.

Mellinger, Philip, "'The Men Have All Become Organizers': Labor Conflict and Unionization in the Mexican Mining Communities of Arizona, 1900-1915." *Western Historical Quarterly* 23 (August 1992): 323-47.

Monkkonen, Eric H., "Introduction." In *Walking to Work: Tramps in America, 1790-1935*, edited by Eric H. Monkkonen, 1-17. Lincoln: Univ. of Nebraska Press, 1984.

———. "Regional Dimensions of Tramping, North and South: 1880-1910." In *Walking to Work: Tramps in America, 1790-1935*, edited by Eric H. Monkkonen, 189-211. Lincoln: Univ. of Nebraska Press, 1984.

———. "Afterword." In *Walking to Work: Tramps in America, 1790-1935*, edited by Eric H. Monkkonen, 235-47. Lincoln: Univ. of Nebraska Press, 1984.

Mullen, Michael. "No Time to Quibble: The Jones Conspiracy Trial of 1917." *Chronicles of Oklahoma* 59 (spring 1982): 224-36.

Preston, William, Jr. "Shall This Be All? U.S. Historians versus William D. Haywood et al." *Labor History* 12 (summer 1971): 435-53.

Rosen, Ellen I. "Socialism in Oklahoma: A Theoretical Overview." *Politics and Society* 8(1), (1978): 109-29.

Rowley, William D. "The West as Laboratory and Mirror of Reform." In *The Twentieth Century West: Historical Interpretations*, edited by Gerald D. Nash and Richard Etulain, 339-57. Albuquerque: Univ. of New Mexico Press, 1989.

Schneider, John C. "Tramping Workers, 1890-1920: A Subcultural View." In *Walking to Work: Tramps in America, 1790-1935*, edited by Eric H. Monkkonen, 212-34. Lincoln: Univ. of Nebraska Press, 1984.

Schofield, Ann. "Rebel Girls and Union Maids: The Woman Question in the Journals of the AFL and IWW, 1905-1920." *Feminist Studies* 9 (summer 1983): 335-58.

Schwantes, Carlos A. "The Concept of the Wageworkers' Frontier: A Framework for Future Research." *Western Historical Quarterly* 18 (January 1987): 39-56.

Sewell, Steve. "Amongst the Damp: The Dangerous Profession of Coal Mining in Oklahoma, 1870-1935." *Chronicles of Oklahoma* 70 (spring 1992): 66-83.

Steffen, Jerome O. "Stages of Development in Oklahoma History." In *Oklahoma: New Views of the Forty-Sixth State*, edited by Anne Hodges Morgan and H. Wayne Morgan, 3-30. Norman: Univ. of Oklahoma Press, 1982.

Taft, Philip. "The IWW in the Grain Belt." *Labor History* 1 (winter 1960): 53-68.

———. "The Federal Trials of the IWW." *Labor History* 3 (winter 1962): 57-91.

Tyler, Robert L. "The IWW and the Brainworkers." *American Quarterly* 15 (spring 1963): 41-51.

Way, Peter. "Evil Humors and Ardent Spirits: The Rough Culture of Canal Construction Laborers." *Journal of American History* 79 (March 1993): 1397-1428.

Weisberger, Bernard A. "Here Come the Wobblies!" *American Heritage* 18 (June 1967): 30-35, 87-93.

White, Earl Bruce. "*The United States v. C. W. Anderson et al.*: The Wichita Case, 1917-1919." In *At the Point of Production: The Local History of the I.W.W.*, edited by Joseph Conlin, 143-64. Westport, Conn.: Greenwood Press, 1981.

———. "The IWW and the Mid-Continent Oil Field." In *Labor in the Southwest: The First One Hundred Years* edited by James C. Foster, 65-80. Tucson: Univ. of Arizona Press, 1982.

Index

Ada, Okla., 24
African-Americans, 19, 20, 32, 52-53, 63, 144, 164, 212n, 225n, 237n
Agents, federal, 105, 111, 115
Agribusiness, 5, 33, 179, 183, 186
Agricultural ladder, myth of, 79
Agricultural Workers Industrial Union (AWIU). *See* Agricultural Workers Organization (AWO)
Agricultural Workers Organization (AWO), 4, 35, 41, 44, 48-50, 52-55, 57, 71-73, 75, 83, 88-89, 93, 95, 103, 115, 136, 144-47, 149-50, 152-53, 156, 161, 166, 174-77, 179, 181-85, 187, 189, 198n, 243n, 246-47n, 253n, 255n; becomes AWIU, 73, 98; condemns violence, 247; 4-in-1 Drive of AWIU, 1924, 153; General Organizing Committee, 137, 146; 1923 Oklahoma City convention of AWIU, 149-51; rejects affiliation with Working Class Union, 88
"Akbar," IWW writer, 182
Alcohol, alcoholism, 43, 58, 60, 62, 213-14n, 217n
Aldridge, George, 136, 139, 146, 243-44n
Alfalfa County, Okla., 149
Alienation, 8

Allied Industries of Oklahoma, 191-92
Alva, Okla., 47, 99, 147-48, 150, 181-82
Amalgamated Meat Cutters and Butcher Workmen (AFL), 110, 156, 248n
American Civil Liberties Union (ACLU). *See also* National Civil Liberties Bureau (NCLB), 161
American Federation of Labor (AFL), 13, 15, 18, 20-25, 28, 35, 42, 55, 63, 66-68, 70, 72-73, 93-94, 101, 103, 107, 109-10, 112, 155-57, 174, 182, 189-90, 202-203n, 205n, 232n, 249n; Bakery and Confectionery Workers Union, 25; boilermakers union, 159; business unionism of, 18; oil workers union, 157
American Labor Union (ALU), 9, 16, 24
American Labor Union Journal, 16
American Legion, 152
American Protective League, 96-97, 116, 164, 219n, 228n
American Railway Union (ARU), 9, 27, 205n
American Southwest, 13
American West, West Coast, Far West, 4, 7, 21, 48, 52, 62, 133, 188; West as industrial frontier, 193

Ameringer, Oscar, 20-21, 27, 157, 207-208n
Amnesties, conditional, 116
Anarchism, anarchists, 102, 123
Anderson, Charles W., 103-104, 109, 115, 231n, 244n
Anderson, Edward E., 160
Anti-Catholicism, 170
Anti-Communist League of America, 192
Anti-semitism, 170
Anti-trust laws, 138
Antiwar movement, 102
Archibald, Dave, 195
Ardmore, Okla., 60, 89, 105, 182, 247n
Arizona, 99, 102, 103, 124
Arkansas, 16, 41, 57, 84, 98, 129, 157-58, 165, 167, 186, 224n
Arkansas City, Ark., 105
Armed robbery, or hijacking, 37, 43-44, 53, 61, 113, 211n
Arnold, Grace, 109
Artisans, 36
Ary, Jack, 99
Ashton, W. G., 47-48, 53, 101
Atchison, Topeka, and Santa Fe Railroad, 111
Atoka, Okla., 166
Augusta, Kans., 72-74, 104-105, 109, 111
Australia, 133
Automobile, "tin lizzie," or flivver, 180-81

Bald Knob, Ark., 166
Baldwin Locomotive Works, 122
Ballew, Bud, 60
Barr, Albert "Doctor," 111, 115, 164, 186; author of *Let Tomorrow Come*, 186
Bartlesville, Okla., 122, 145, 154, 171, 202n, 245n
Beckham County Advocate, 24
Bedmarcek, Paul, murder of, 149

Belgium, 108
Benson, George, 129-31
Berg, Arthur, 167-70, 172-73; as "Dug Kugher from Everywhere," 167
Berger, Victor, 20, 24, 207n
Bill of Rights, 134
Binders and headers, 36-37
Bindlestiffs. See Hoboes, 38, 40, 211n
Bing Alexander, 99
Bisbee, Ariz., 99, 113, 132, 226n, 229n
Blackwell, Okla., 136
Blake, Aldrich, 164
Blumberg, A. M., 114-15, 255n
Bobba, R. J., 103, 113
Bolshevik Revolution, bolshevism, 110, 127, 141
Bonanza farms, 36
Bondhauer, Katherine, firing of, 119
Bonstein, George, 125, 239n, 243n
Boom towns, 11, 32-33, 57-58, 60-61, 71, 75
Boose, Arthur F., 72-73, 95, 103, 113, 186, 231n, 236n
Bootleggers, bootlegging, 43-44, 119, 152
Boston, Mass., 127
Boyd, E. M., 106, 108-109, 111, 115, 128, 130, 232n
Boys' Working Reserve Force, 144
Bradley, Henry, 156
Brewster, A. C., 169-70, 173
Brink, E. W., 69
Bristow, Okla., 122, 130, 179
British Isles, 9, 41
Brooklyn, N.Y., 145, 193
Brotherhood of Timber Workers (BTW), 40, 67, 74, 81-82, 84, 94, 190
Bugher, William, 166
Bulgarians. See Europeans
"Bummery," the, 29, 194, 257n

INDEX

Bureau of Investigation, U.S. Department of Justice, 96–97, 102
Bureau of Labor Division of Information, United States, 46
Burns detective agency, 97
Buse, Mrs. Elmer F., 20
Businessmen, 76, 93, 110, 117, 144
Businessmen's frontier, 199n
Business progressives, 6, 79, 116–17, 134, 141, 199n
Butler County, Kans., 232n
Butte, Mont., 31, 101

Caldwell, Kans., 136
California, 46, 63, 66–67, 150, 155, 160, 166, 189, 250n, 256n
Camp Funston, 136
Canada, 36, 41, 125, 180; British Columbia, 185; Manitoba, 36, 182; Saskatchewan, 36
Canadian County, Okla., 168
Cannery and Agricultural Workers Industrial Union (CP), 189, 256n
Cargill, O. A., 151
Carney, John J., 91, 148, 152, 165, 167–69, 172–73
Carpenter, Ethel E., 19–20, 33, 39–40, 82
Carpenter, J. D., 20
Carpenters and joiners, 18
Carroll, Dill, 9
Carter Oil Company, 59, 97, 105, 123–24, 126, 128–29, 131, 141–42, 154, 241n
Casebolt, William, 39
Casey, Harry "Nuff Said," 112, 122, 124, 238n
Casper, Wyo., 160, 178
Centralia, Wash., 150
Chaffin, C. H., 192
Chambers of commerce, 74, 119, 191
Champlin Petroleum Company, 191

Chaplin, Ralph, 38, 100, 161, 191, 233n
Cherokee, Okla., 30, 47, 58, 98, 149
Cherokee Strip Livestock Association, 6
Chicago, Ill., 52, 90, 100, 102, 139, 176, 178, 185, 206n
Chickasha, Okla., 112, 122
"Children's Crusade," 115
Choctaw beer, 60
Choctaw Nation, 7; guest workers of, 9
Christiansen, Otto, 115
Churches of Christ, Federated Council of, 88
Cities Services Petroleum Company, 191
Citron, Oscar, 171
Civil War, 7, 36, 86–87, 225n; Confederacy, 86
Clan of Toil, 81
Claremore, Okla., 132
Clark, Stanley J., 24, 28, 103, 113, 189
Clayton Anti Trust Act, 135
Cleveland, Ohio, 206n
Cleveland, Okla., 145, 154
Cleveland County, Okla., 90–91, 111
Clinton, Charles, 69
Clinton, Okla., 180
Coalgate, Okla., 9, 22
Coal miners, 8, 12, 16, 154, 166, 182, 189–90
Cole, Redmond S., 87, 126, 128, 224n, 234–36n, 239n, 241–42n
Colescott, Ralph, 177
College students, 41–42, 180
Colorado, 30, 89, 182, 248n
Combine-harvester, 163, 179–80, 183, 188
Commerce, Okla., 171
Committee on Public Information, 93

Commons, John R., 26
Communist Party of America
 (CP), 174, 185, 187, 189, 192-93
Company unions, 154, 182, 190,
 248n
Congdon, Glenn, 102, 107, 223n,
 233n
Congress, United States, 86-87,
 96, 135, 193; Senate, 101
Congress of Industrial
 Organizations (CIO), 22, 162,
 174, 179, 186, 189, 191-92,
 194-95, 249n, 256-57n; in
 Oklahoma, 190
Conner, Tom, 44
Conscription, 77, 86, 90
Conscription (Draft) Act, 74, 87,
 91, 95
Construction workers, 11, 67, 100,
 174
Construction Workers Industrial
 Union (CWIU), 63, 103, 153,
 182
Cook, Fred, 97
Coolidge, Calvin, 116
Corn Belt, 150
Cotteral, John H., 91
Cotton, 7, 79; price collapse, 85
Councils of Defense, 94-96,
 140-41; Oklahoma Council of
 Defense, 116-17, 164; Seminole
 County Council of Defense, 134
Cox, Mrs. Allie L. (friend of Frank
 Little), 31
Coxey's Army, 44
Craft unions, 9, 21
Creek County, Okla., 58, 61, 66,
 99, 101, 109
Criminal syndicalism, 12, 120,
 133-34, 137, 139-42, 146, 151,
 161, 163, 166-69, 183-84, 192,
 195; Oklahoma law, 133;
 Oregon law, 135
Crowe, V. P., 152
Cullison, J. B., 138

Cumbie, J. T. "Tad," 28, 89
Cushing, Okla., 3, 57, 59, 70, 72,
 109, 111, 166, 190-91, 232n
Cushing oilfield, 58, 221n, 231n

Daily Oklahoman, 40-41, 47, 53, 88,
 101, 110, 119, 151, 193
Dakotas, the, 36, 104, 180
Davis, A. B., 24
Debs, Eugene V., 9, 16, 27
Deep Rock Petroleum Company, 191
Delaney, T. F., 23
DeLeon, Daniel, 28, 89
Democratic party, 6, 94, 126, 134,
 141, 157, 167-68, 206n, 225n
DeNoya (Whizzbang), Okla.,
 159-60, 249n
Denver, Colo., 20
Department of Agriculture, United
 States (USDA), 37
Department of Justice, United
 States, 91, 113, 172, 219n
Department of Labor, Oklahoma,
 30, 101, 145, 151, 180
Department of Labor, United
 States, 46
Detroit, Mich., 125
Dewar, Okla., 90, 99
Dimmit, William, 155
Dissent, limits of, 120
"Doc" (IWW writer), 136, 154
Doree, E. F., 83, 231n
Dougherty, T. G. F., 83
Doyle, Thomas H. (Okla. appeals
 court judge), 71
Doyle, Tom (AWIU secretary-
 treasurer), 144, 177, 253
Drug abuse, 43, 58, 62
Drumright, Okla., 3, 31, 49, 54,
 60, 66-67, 69-70, 72-74, 95,
 99, 101, 109-10, 112, 127, 145,
 154, 165, 176, 190, 191-92; 1919
 telephone operators strike, 127,
 134; Patriotic Committee of
 Drumright Citizens, 99

INDEX 287

Drumright Daily News, 111
Drumright Evening Derrick, 99, 110
Dual unionism, 22, 32
Duffy, James, 72
Duluth, Minn., 150
Durant, Okla., 17, 119, 202

East Coast, United States, 4
Eastman, Phineas, 67, 74, 101, 103, 111-12, 115
Edgecombe, Fred, 112, 235n
Edwards, Forrest, 54, 95, 103, 253n
Edwards, Thomas A. (Okla. appeals court judge), 173
Eighth Circuit Court of Appeals, United States, 115
El Dorado, Ark., 165
El Dorado, Kans., 111
Ellis, Kans., 52
El Reno, Okla., 89
Embree, A. S., 182
Empire Oil Company, 166, 257n
Englehardt, Martin, 120
Enid, Okla., 36, 41, 47-55, 90-91, 98, 133, 136-37, 145-49, 151-53, 161, 253n
Enid Daily Eagle, 41, 44, 138, 243n, 246n
Enid Daily News, 139, 243n
Enid Morning News, 138
Espionage Act, 87, 91, 95-96, 103, 114, 228n
Ettor, Joseph, 125, 187
Europe, Europeans; 4, 71; Bulgarians, 68, 73; Eastern Europeans, 73, 171; Finnish iron miners, 72; France, 98; Italians, 171
Evangelical style of IWW, 78, 81, 187
Evans, R. E., 155
Evans, T. D., 106-107, 109, 233n
Everett, Wash., 125

Fain, John, 91
Fargo, N.D., 137, 153
Farm Bureau, 94
Farmer-Labor Reconstruction League, 150, 166, 256n
Farmers and Laborers Protective Association (Texas), 80, 88, 90
Farmers Educational And Labor Union, (Farmers Union), 50, 53
Farmers Emancipation League, 80
Farms, farmers, 37-42, 47, 49, 52-54, 58, 79-80, 82, 83, 85, 92, 98, 116-17, 136, 144-46, 148, 151, 153, 166, 179-81
Federal conspiracy trials of IWW. *See* Industrial Workers of the World
Federal Labor Unions of AFL (FLU), 42, 202n
"Fellow Workers" (IWW pamphlet), 157
Fenton, Edgar, 190, 256n
Fenton, George, 67
Findlay, James G., 238n
Findley, Charles, 101
Finley, Ira, 24, 189
First Amendment, 193, 237n
Fisher, Joe, 144, 253n
Five Tribes, 7
Floaters. *See* Hoboes, tramps, and bums
Flying squads, 54, 155
Foodstuff Workers Industrial Union (IWW), 156
Foote, E. J., 25, 28-29, 33, 207n
Forberg, Lillian, 29
Fort Smith, Ark., 188
Foster, J. T., 123-24, 237n
Fourteenth Amendment, 134
Fox, Mat K., 136, 145
Foy, James, 62
Francik, Wencil, 99, 104, 109, 111, 114-16, 123, 176, 186
Fraser, Ted, 72

Free Employment Bureau, Okla., 50, 98, 136, 145-46, 149, 180
Freeman, Fred "Uncle Fred," 84
Free speech fights, soapboxing, 6, 12, 50, 68-69, 95, 193; in Fresno, 30; in Kansas City, Mo., 31, 125; in San Diego, Calif., 125; in Spokane, Wash., 24, 30, 125, 207n; in Tulsa and Muskogee, 24, 67-69
French, Joe, 107-108, 232-33n
Frisco Railroad. *See* Saint Louis and San Francisco Railroad

Gallagher, Frank J., 100, 103-105, 111, 115-16, 153, 176, 186, 236n, 253n
Gamblers, gambling, 43-45, 52, 58, 60
Garfield County, Okla., 36, 51, 137, 149, 152, 181
Gaylord, E. K., 193
Geary, Okla., 24
German, Germany, 78, 88, 92-94, 96, 100-102, 107, 119, 141, 168, 232n; alleged German agents, 102; German prisoners of war, 98
Gibson, Charles, 178
Giovannitti, Arturo, 125, 237n
Glen Pool oilfield, 58-59
Gompers, Samuel, 18, 21, 23
Goose Creek, Tex., 73
Gordon, Oscar E., 104, 115-16, 186
Gore, Thomas (U.S. senator from Oklahoma), 141
Gould, Jay, 26
Grant County, Okla., 147
Grant County Socialist, 28
Great Depression, 186, 188, 193
Great Southwestern Railway Strike of 1886, 26
Green, J. W., 132
Green Corn Rebellion, 3, 4, 11, 76-92, 94, 112, 117, 135, 190, 222n

Gregory, Thomas (U.S. Attorney General), 96
Gresbach, Joseph, 111, 114-15, 255n
Griffeath, Ernest, 83
Griffin, David and Sadie, 124, 129
Gross, Marion C., 137-38
Gulf Coast oilfields, 67, 103, 231n
Gulf Oil Company, 59, 176
Guthrie, Okla., 16, 31, 41
Gypsy Oil Company, 100

Hagerty, Father Thomas J., 16, 25, 30, 84, 202-203n, 224n
Hailleyville, Okla., 167, 169
Halcrow, Tom, 31
Hall, Covington, 31, 40, 82-84, 190
Hall, John, 123-24, 129-32, 238n
Haney, James, 182
Hanson, Nils, 41
Harding, Warren Gameliel, 176
Harlow, Victor, 86, 141
Harlow's Weekly, 86, 102
Harper, George, 124, 127, 129, 131
Harper County, Okla., 119
Harreld, John, 115-16
Harrison, Luther, 134, 135, 193
Hartshorne, Okla., 154
Harvest working conditions, 38, 40
Haskell, Charles, 60
Hathaway, M. A., 95
Haywood, William, 15, 19-20, 25, 27-28, 33, 35, 54, 57, 73-74, 100, 132, 161, 175, 208n, 215n; Sp. recall of, 25
Healdton, Okla., 60, 72, 77, 119, 191, 232n
Healdton Pool, 58, 89
Healy, J. M., 66
Henryetta, Okla., 90, 105, 111, 171, 202n
Hispanics, 63, 73; in IWW, 66
Hoboes, tramps, and bums, 11, 12, 20, 30, 33, 36-42, 47, 50, 52-55, 57, 62-63, 75, 81, 104,

INDEX 289

144-46, 148, 150-51, 179-80, 193-94, 253n; hoboes as franc tireurs, 35, 40, 194; "hobo jungles," 37, 44, 50, 51, 53, 99, 145, 181; "hobo's delight," 38; "honk honk hoboes," 181; middle-class fears of hoboes, 44; origin of term from "hoe boy," 40; strikes by, 45
"Hobo Hollow," Ark., 84
Home Guard, Okla., 95; Tulsa, 105, 109
Hominy, Okla., 100, 160
Homosexuality, 43-44, 62, 213-14n
Howe, Thomas J., DOJ agent, 115
Hughes, Forrest (Oklahoma City city attorney), 157
Hughes County, Okla., 85
Humboldt, Ariz., 131

Idaho, 20, 62, 134
Illinois, 30, 41
Immigrants, 164
Imperial Nighthawk, 165
Incorporation process, 93, 116, 117, 120, 184, 197n
Indiana, 164
Indian Territory (IT), 6-8, 10, 16-17, 199n, 202n
Indian Territory Illuminating Oil Company, 154, 191
Industrial Pioneer, 153-54, 163, 186
Industrial Union Bulletin, 8, 208n
Industrial unionism, 9, 21, 23, 40, 42, 54, 78, 83, 184, 189-90, 193, 249n
Industrial Union Manifesto, 16
Industrial Union News, 30
Industrial Worker, 18, 20, 23, 25, 40, 71, 82, 100, 186-87, 207-208n, 233n
Industrial Workers of the World (IWW), 3, 11-13, 15-26, 28, 31-33, 35-36, 39-40, 42, 44, 46-53, 55-58, 60, 63, 66, 68-69, 71, 72, 74-79, 81-84, 86, 88, 91-95, 97-105, 109-12, 115-19, 121-23, 126-28, 131-33, 135, 137-48, 152, 154-56, 160-65, 167-71, 173-75, 177, 180-192, 194-95, 197-98n, 203n, 206, 213n, 222n, 232n, 237n, 244n, 247-48n, 255n; centralizers vs. decentralizers, 174-75, 183; Detroit faction. *See* Workers International Industrial Union (WIIU); dispute over political prisoners and conditional amnesties, 179; Education Bureau, 177, 250n; Emergency Program faction (EP), 173, 175, 178; federal conspiracy trials of IWW: Chicago trial, 3, 24, 58-59, 87, 91, 103, 111-12, 114-15, 121, 125-26, 225-26n, 237n, Sacramento trial, 121, 236n, Wichita and Kansas City trials, 3, 91, 118, 122-25, 127, 132, 153, 181; "Four Treys" faction, 175; function of union halls, 220n; General Defense Committee (GDC), 120, 121, 133, 174; General Executive Board (GEB), 29-30, 67, 70, 72-75, 91, 95, 100, 132, 155, 174, 177-79, 182, 184-85, 253n; General Recruiting Union (GRU), 179; Kansas City local, 45; mining department, 66; Muskogee IWW Women's Auxiliary, 19; Oklahoma City Excavators Industrial Union, 19; Oklahoma City Industrial Union No. 239, 17, 29; Railroad Workers Industrial Union, 253n; refusal to sign

Industrial Workers (*continued*)
contracts, 187, 193; rejection of ballot box by, 5, 10, 18, 20, 24, 30–32; rejection of political action, 193; rejection of violence, 75; schisms: Sherman-DeLeon-Trautmann split, 28, 208n, DeLeon-Bummery split, 29, IWW 1924 split, 163, 174–75, 183, 186; Shawnee Mixed Industrial Union, 17; term limits for officers, 183; Tulsa IWW headquarters, 105
International Association of Machinists (IAM), 18, 23, 32, 122
International Association of Oil Field, Gas Well, and Refinery Workers (AFL), 103, 155, 190
International Brotherhood of Oil and Gas Well Workers (AFL), 66–67
International Typographical Union, 22
Iowa, 46, 104, 139, 186
Isenhour, Clure, 91
Isenhour, Obe, 91

Jaffe, Eli, 193
Jenkins, Tom J., 130
Jerome, Ariz., 113, 229n
Jingoism, 100–101
Job delegate system, 36, 44, 48–50, 52–53, 55, 57, 72, 98, 105, 120, 136, 144–46, 148, 154, 157–58, 161–62
Job sharks, 47, 203n
Johnson, Gunnard "Bernard," 107, 232n
Joliet, Ill., 28
Jones, "Mother" Mary, 16
Jones Family, the, 86, 90–91, 126, 224n
Joplin, Mo., 17, 171

Kansas, 3, 16–17, 39, 41, 45–52, 56, 67, 73, 84, 88, 98, 100, 104, 110–12, 125, 144–45, 147, 150, 153, 158, 160, 168, 171, 178, 180; industrial court, 150; jails, 114
Kansas City, Kans., 114
Kansas City, Mo., 7, 30–31, 45–46, 48, 99, 136, 139, 229n; 1916 farm labor conference at, 53
Katz, Michael, 29
Kay County, 152
Kerles, Walter, 130
Kerns, E. C. (Cushing Klan "cyclops,"), 166
Kiefer, Okla., 60–61, 67
Kight, H. Tom (state representative), 192
King, Charles, 60
King, W. C., 49
Kirby, John, 94, 227n
Kirby Lumber, 82
Kirk and Gustafson detective agency, 97, 106
Knights of Labor, 6, 9–10, 41, 81–82, 84, 198n, 207n; role in 1894 Coal Strike, 10
Knights of Liberty, 3, 108–10, 164, 233n
Koen, James, 57
Krebs, Okla., 9–10
Krieger, Charles, 112, 122–25, 127, 129–33, 140–42, 186
Ku Klux Klan (KKK), 63, 108, 110, 119, 152, 161, 163–67, 170–71, 183, 246n
Kusa, Okla., 90

Labor, labor unions, 7, 18, 120, 170; injunctions against, 166; labor unrest, 58; spies among, 120
"Labor fakirs," derogatory for union officers, 175

Index

LaFallette, F. W., 108
Lamont, Okla., 147
Landis, Kenesaw Mountain, 112
Land League of Oklahoma and Texas, 31, 80
Land prices, 85; fraud in, 79; speculation in, 79
Langston, J. Luther, 22, 24, 190, 256n
Latchem, E. W., 49, 51, 144, 161, 185–86, 253n
Law, Jack, 49, 67, 103, 121
Law as superstructure in society, 121
Lawrence, Mass., 4, 6, 35, 121
Lawton, Okla., 122, 246n
Leavenworth, Kans., 77, 88, 115, 125, 129, 131, 154–55, 185, 187
Leen, John J. "Gasoline Slim," 163
LeFevre, Wells, 84, 224n
LeFlore County, Okla., 188, 205n
Lehigh, Okla., 9, 154
Lenopah, Okla., 30
Lester, E. F., 168
Lever Act, 95, 114
Liberty Bonds, 107, 131, 140
Lincoln, Ora, 152
Little, Alonzo, 30
Little, Frank, 20, 25, 30, 31, 33, 48, 67, 81, 95, 100–101, 209n; murder of 30–31, 101
Little River, 77, 185
Lorton, Eugene, 102, 241n
Los Angeles, Calif., 139
Louisiana, 16, 31, 40, 67, 73, 80, 84, 98, 103, 157–58, 168
Lowe, Caroline A., 115, 125–27, 130, 240n
Lucas, E. L., 105, 108, 233n
Lumber industry, 35, 75
Lumberjack, The, 82
Lumberjacks, 30, 58, 99, 151, 174, 252n; lumber camps, 64; unions of, 153

Lyons, Eugene, 123, 128, 130–31, 241n
Lyons, Harry, 122

McAlester, Okla., 77, 167; state prison, 139, 170, 172–73
McCook, John J., 41
McCurry, John (Pinkerton agent in Tulsa), 106, 232n
McGinnis, W. P., 91, 113–14, 236n
McGuire, Peter, 18
Macklin, Chester, 109, 112, 234–35n
McNabb, L. C., 85
Madison Capital-Times (Wisc.), 193
Magnolia Oil, 59
Marine Transport Workers Union (IWW), 151
Marxism, 18
Masculinity, 43, 214n
Mathewson, Stanley, 26
Matteson, E. E., 23
Mead, William, 82
Meatpackers, packinghouses, 17, 156–57, 189, 248n
Mechanization, 4, 5, 143
Medford, Okla., 25, 145–46, 244n
Medford Challenge, 25
Merryville, La., 67, 74
Mesabi Range, Minn., 22, 72
Meserve, John B., 97, 105, 106, 233n
Metal Mine Workers Industrial Union, (IWW), 167, 170
Mexico, Mexicans, 16, 73, 101, 124, 187, 212n, 246n
"M. H.," WIIU member, 30
Miami, Okla., 167
Michigan, University of, 125
Midcontinent Field, 3, 5, 7, 58–59, 103, 113, 154–55, 160, 178, 191
Mid-Continent Oil and Gas Association, 120
Mid-Continent Refinery, 192

Military intelligence, 120
Miller, Jack, 38, 54
Milner, H. Grady, 24
Miners, hardrock or metal, 5, 7–8, 11, 15, 30, 87, 90, 170–71, 182, 190
Minneapolis, Minn., 52, 103–104
Minnesota, 103, 124, 133
Mississippi, 74, 80, 84, 150, 223n
Missouri, 17, 41, 46, 99, 126, 129, 170, 180
Missouri, University of, 126
Modernization, 12
Moffet, Okla., 86
Montana, 99, 124, 133
Montgomery, W. B., 123
Moore, Fred, 115, 125–28, 130–32, 137–39, 242–43n
Moore, Okla., 111
Moran, John, 109, 232
Moran, W. H., 137, 139
Morgan, J. P., 87
Morris and Company (meatpackers), 150
Morris[on], E. G., 106, 232n
Moss, Austin Flint, 126, 128, 130–31, 239n
Mount Copper Boiler and Iron Works, 122, 130
Moyer, Charles, 19
Mulkes, Harold O., 132–33, 148, 165
Munroe, Thomas L., 128
Munson, Henry H. "Rube," 87–91, 103, 112, 225n
Murphy, Frank, 9
Murphy, Joseph, 187, 211n
Muskogee, Okla., 17–19, 69, 123–25, 129, 154, 167, 188, 205n; Citizens' Alliance, 17
Muskogee County, Okla., 86
Muskogee Daily Phoenix, 101

Nagle, Patrick, 91
National Association of Manufacturers, 191, 205n, 227n
National Civil Liberties Bureau (NCLB), 108–10
National Defense Act, 96
National Farm Labor Exchange, 46
National Guard, Mo., 99; Oklahoma, 127, 134, 141, 156, 192–93
National Industrial Recovery Act, 191; Section 7a of NIRA, 191, 193, 248n, 257n
National Labor Relations Act (Wagner Act), 191, 193, 257n
Native Americans, 6, 19, 102
Nativism, 19, 182, 190
Nebraska, 46, 84, 104, 144, 150
Nef, Walter, 50, 54
New Century, 24
New Mexico, 100
New Rockford, N.D., 156
New unionism, 182
New Wilson, Okla., 72, 89
New York City, 27, 255n
Nightriding, 86, 225n
Nilsson, B. E., 83
Non-Partisan League, 98, 190
Norman, Okla., 11, 185
Norman Political Confederation, 185
North Dakota, 98, 150
Northern Plains, 7
Nowata County, Okla., 30
Nowlin, Dave, 169, 173

O'Donnel, William, 28
Ohio, 60, 67, 137, 253n
Oil, Gas, and Petroleum Workers Industrial Union No. 230 (IWW). *See* Oil Workers Industrial Union
Oil and Gas Journal, 72, 110, 165
Oil field workers, 11–12, 30–31, 33, 49, 54, 57–58, 61–70, 73–75, 80–81, 100, 104–105,

INDEX 293

110-11, 153-55, 158-60, 161-62, 166, 175, 179, 184, 190, 194, 247n
Oil field working conditions, 61-62, 64, 157, 176
"Oil Workers!" (IWW pamphlet), 157, 162
Oil Workers Industrial Union, 3, 49, 58, 64, 69-70, 73-74, 89, 91, 100-106, 111, 115, 126, 144, 149, 152-62, 165-67, 173, 175-78, 182, 184, 249n, 253n; finances, 176; general organizing commitee, 156, 158, 248n; 1923 convention in Oklahoma City, 176; Oklahoma City headquarters, 156, 176-77
Oil Workers International Union (OWIU-CIO), 179, 191-92
Oklahoma, 3, 7, 9, 11-13, 16, 19, 22, 25, 28, 30-32, 36, 39-41, 45-47, 49-51, 55-58, 61, 67, 72, 76, 79, 80, 84-85, 93, 97-100, 103-105, 110, 112, 115, 117, 122, 135, 140, 144, 147, 150, 152-55, 157-58, 160-61, 163, 165, 168, 170, 176, 178-80, 182-85, 191, 194, 223n, 246n, 256n; 1889 Land Run, 30; prohibition laws, 43-44; state legislature, 59, 120
Oklahoma, University of, 185
Oklahoma Agricultural and Mechanical College, 34
Oklahoma City, Okla., 16, 19, 20, 23-24, 28-29, 31, 33, 70-71, 95, 101, 103, 105, 110, 114, 136, 149, 151, 157-58, 162, 164-66, 176, 178-79, 189, 192, 237n, 246n; 1911 streetcar strike in, 23; 1921 packinghouse strike in, 156-57, 248
Oklahoma City Printing Trades Council, 22
Oklahoma City Times, 193

Oklahoma Criminal Court Appeals, 139, 170, 173
Oklahoma Criminal Syndicalism Act. *See* Criminal syndicalism
Oklahoma Federationist, 110
Oklahoma Labor, 190
Oklahoma Labor Unit, 22, 71, 203n
Oklahoma Leader, 126, 157
Oklahoma News, 101, 157, 248n
Oklahoma Pioneer, 27
Oklahoma Pipeline Company, 62
Oklahoma Press Association, 101
Oklahoma Renters Union, 31, 80, 85-86
Oklahoma State Federation of Labor (OSFL), 10-11, 22, 30, 32, 42, 110, 135, 166, 190, 205-206n, 226n; as Twin Territories Federation of Labor, 10
Oklahoma Territory (OT), 6-7, 9-10, 16, 17, 24, 46
Okmulgee, Okla., 110, 163
Oliphant, L. W., 112
Omaha, Nebr., 91, 156, 185, 205n
One Big Union concept, 162, 174
One Big Union Monthly, 135, 137
101 Ranch, 36
O'Neil, Patrick, "Uncle Pat," 16
Open shop movement, 17-18, 182
Oregon, 150
Osage County, 101, 154
Osage Hills, 3, 108
Ottawa County, Okla., 172
Owens, George E., 27
Owens, L. G., 128, 131
Owensboro, Ky., 124

Pacific Northwest, 29, 35, 46, 64, 89, 99, 182, 186
Palmer, A. Mitchell, 127
Panic of 1907, 29, 33, 67, 212n
Parker, J. R., 98
Parker, W. R., 144
Paterson, N.J., silk strike, 4, 35, 155

Pawhuska, Okla., 145
Pawnee, Okla., 109, 126, 128
Payne County, Okla., 30
Pennsylvania, 9, 67
Pennsylvania State University, 163
Pentecostalism, 81
People's College, Fort Scott, Kans., 125
Perry, Okla., 17, 101, 122, 130
Peterson, Nels, 39
Petroleum, petroleum industry, 13, 33, 56, 59, 74, 94, 102, 143, 176, 183-84, 194
Petroleum Labor Policy Board, 191
Pettibone, George, 19
Pew, J. Edgar, 105, 108, 122, 124, 129, 131-32, 233n, 255n; bombing of Pew home, 106-107, 112, 122-23, 126, 129-30, 141, 232n
Philadelphia, 112, 186
Phillips Petroleum Company, 154
Picher, Okla., 170-71
"Pie-cards" (derogatory term for union officers), 175
Piece-rates, 159, 160, 162
"Piece Work and the Tank Builders" (IWW pamphlet), 162
Pinkerton detective agency, 97, 124, 129, 233
Pittsburg, Kans., 16
Pittsburg County, Okla., 167
PM (New York newspaper), 193
Ponca City, Okla., 36, 136, 152
Pontotoc County, Okla., 77, 85-86, 111
Populist Party, populism, populists, 82, 84, 94, 116, 164
Porter, Claude, 114
Portland, Ore., 175
Pottawatomie County, Okla., 90-91
Powers, W, J., 105, 231n
Prairie Oil and Pipeline Company, 59, 70, 73, 97
Prescott, Ariz., 122

Profintern (CP), 185
Progressive Bookstore, 193
Progressivism, Progressive thought, 46, 93, 95
Prostitution, prostitutes, 43, 58, 60, 119
Pryor, Tom (alleged killer of Paul Bedmarcek), 149
Pullman Strike, 1894, 26
Purdy, Victor, 22, 190, 256n
Pure Oil Company, 192

Quakers, Quakerism, 30
Quapaw, Okla., 171

Railroads, 7-8, 19, 36, 47, 51, 143, 151, 153, 181; railroad agents, 47; railroad workers, 252n
Rebellion, 84
Red Scare, 114, 119
Reeder, Walter M., 87, 89-91, 103, 112, 226n
Republican party, 102
Rice, A. A., 69
Richardson, Charles, 106
Ringling, Okla., 60
Roberts, Jasper, 24-25
Robertson, Fred, 114
Robertson, James Brooks Ayers, 114, 127, 134, 156; declares martial law in coal fields, 127
Robinson, J. W., 124
Rockefeller, John D., 129
Rock Island Railroad, 148, 167
Rockwell, A. W., 67
Rocky Mountains, 9, 15
Rogers, Bruce, 24
Rogers County, Okla., 132, 192
Roosevelt, Franklin Delano, 191, 193
Rowan, James, 174-75
Royal Dutch Shell, 59, 73, 192
Russia, Soviet Union, 104, 110, 175, 221n
Ryan, J. F. "Frank," 24, 69, 100-101, 107, 232n

INDEX 295

"S." (Oklahoma IWW member), 18
Sabine River, 67
Sabotage, 6, 10, 18, 21, 25, 26, 28, 52, 54, 94, 97, 102, 119, 127, 134, 167, 193; ca'canny, rattening, playing the Hoosier, skulking, soldiering, 26; origin of term from *sabots*, 25; "sabotabby cat" as symbol of, 102; strike on the job, 26, 36
Sacco and Vanzetti, 125
St. John, Vincent, 28, 30
Saint Louis, Mo., 7, 45, 111, 122
Saint Louis Post-Dispatch, 193
Sallisaw, Okla., 85
Sand Springs, Okla., 109, 120, 127, 129, 171
San Francisco and St. Louis Railroad, 11
San Pedro, Calif., 178
San Quentin Penitentiary, Calif., 139, 244n
Saposs, David, 161
Sapper, Michael, 104, 109, 111, 114-15, 234-35n
Sapulpa, Okla., 110, 185
Sasakwa, Okla., 90
Schnell, Carl, 111, 115
Schwartz, Nick, 158; defects to Emergency Program, 178
Scientific management, 18, 159, 162
Scissorbills, 42, 213n
Seattle, Wash., 54, 128
Seattle General Strike, 125, 127, 135
Seminole, Okla., 191, 257n
Seminole County, Okla., 77, 85
Seneca, Mo., 87
Sequoyah County, Okla., 85-86, 188
Shamrock, Okla., 166
Shaw, Alan, 193
Shawnee, Okla., 17, 18, 23

Sheldon, L. N., 190, 256n
Sherman, Charles O., 28-29
Shipman, W. G., 130
Shirey, John, 91
Short, George F., 139
Shreveport, La., 165
"Silent agitators," or stickerettes, 102, 145, 230
Simonton, Clinton, 29
Simpson, D. N., 136
Sinclair, Harry F., 71
Sinclair, Upton, 60
Sinclair Oil Company, 100-101, 154, 191
Sioux City, Iowa, 139
Skeedee, Okla., 109
Skelly Oil Company, 159
Skiatook, Okla., 111
Smith, C. M., 172
Smith, Ernest J., 137-38
Smith, Walter, 70
Snyder, Okla., 125
Social banditry, 85-86
Social Democrat, 24, 28
Socialist Labor Party (SLA), 16, 25, 28-30
Socialist Party of America (SP), 4, 6, 12, 16, 18, 20, 22-23, 25, 27-28, 31-32, 35, 81, 84-85, 87, 89-90, 94, 103, 110, 113, 115, 117, 125, 128, 135, 138, 140, 141, 158, 165-66, 187-89, 197n, 208n, 224-25n; National Executive Committee, 25-27; "Red" wing versus "Yellow" wing, 24, 27; SP antisabotage amendment, 123, 207, 208n; SP in Oklahoma, 11, 24-25, 27, 32-33, 39, 69, 77, 78, 80, 116, 168; SP in Oklahoma City, 23; SP in Texas, 78, 208n; Women in SP, 208n; Womens National Committee, 125
Socialist Trades and Labor Alliance, 16

Solidarity, Industrial Solidarity, New Solidarity, 31, 44, 45, 49–50, 52, 66, 69, 71, 82, 95, 98, 100, 104, 145, 155, 206n, 230n, 246n; as *Industrial Solidarity*, 147, 149, 152–53, 158, 159–60, 170, 172–73, 181–82, 246n; as *New Solidarity*, 136, 154
"Solidarity Forever" (IWW song), 191
South Canadian River, 90
South Dakota, 46
Southern states, 80
Southern Lumber Operaters Association, 67
Southern Tenant Farmers Union (STFU), 188, 256n
Southwestern Bell Telephone Company, 127
Spears, John, 90
Spence, Homer C., 88–91
Spiro, Okla., 186
Standard Oil, 59–60, 66, 71, 104, 128, 131, 138, 192; Standard of Indiana, 59, 97, 134; Standard Oil of New Jersey, 59, 154, 248n; Standard Oil of New York, 59; Standard Oil of Ohio, 60
Stillwater, Okla., 30
Street, Jane, 20
Strikebreakers, 21, 23, 26, 84, 124, 157, 159, 170; Tri-State miners as strikebreakers, 170
Strong, Walter, 190, 256n
Sun Oil Company, 59, 225n
Supreme Court of the United States, 134
Sweeney, Edward, 99
Syndicalism, 9–10, 13, 18, 31, 78, 80, 118, 201n; general strike as tactic, 162

Taft, Calif., 178
Tahlequah, Okla., 119
Tampico, Mexico, 16
Taylor, Frederick Winslow, 26
Teachers, 119–20
Tenant farmers and sharecroppers, 4, 12–13, 31, 40–42, 76–81, 83, 85–86, 92, 116, 165, 188–89, 194, 201n, 224–25n; crop lien and, 80; diet of, 80
Terrell, Grover Jackson "Jack," 122, 133, 136–42, 145, 243–44n
Texaco, Texas Company, 59, 66, 192
Texas, 16, 24, 62, 67, 84, 86, 103, 124, 153
Texas fever, 85
Thiel Detective Agency, 120
Thompson, E. P., 121
Thompson, Fred, 175, 209n
Threshing, 39, 47, 54
Tidewater Refinery (Drumright), 192
Tonkawa, Okla., 152
Topeka, Kans., 7
Towne, R. H., 125
Townsend, H. H., 97, 108, 124, 233n
Trautmann, William, 28–29, 66
Tri-State mining region, 167, 171, 182
Trotter, Harry, 103, 113, 203n
Tulsa, Okla., 3, 20, 24, 49–50, 58, 60–61, 67–69, 72, 95, 97, 99–102, 105, 107, 109–10, 119, 122–23, 125–26, 131, 165, 217n, 228n, 236n, 239n; chamber of commerce, 108; city commission, 69, 72; 1921 race riot, 106
Tulsa County, Okla., 97, 101, 127–28; jail, 129
Tulsa Democrat, 76, 131
Tulsa Labor Press, 185
Tulsa Outrage, 3, 4, 24, 108–10, 119, 122, 165
Tulsa World, 43, 102, 105–107, 110, 119, 131, 138
Typhoid, 64

INDEX 297

Unemployed Councils (CP), 189
Unemployment, 61, 147, 213n; technological unemployment, 12, 32, 143, 161, 182, 186, 249n
"*Union Scabs—and Others*" (Ameringer), 21
United Brotherhood of Carpenters and Joiners, 18, 32
United Mine Workers of America (UMWA), 9-10, 16, 19, 22-23, 127, 192, 195, 213n; in Kansas, 171; 1919 strike of, 134
United State's Marshal Service, 109
Universal Union, 89-90, 112, 115, 126
Unlawful Associations Act, Australia, 133

Vagrants, vagrancy, 12, 23, 47, 51, 106, 149
Van Buren, Ark., 16, 84
Vanderveer, George, 115, 124
Veterans of Industry of America, 189
Vigilantes, vigilantism, 12, 94, 99, 112, 120, 137, 140, 161, 164, 184; countervigilantism, 79
Voice of Labor, The, 16
Vold, Paul, 137, 139
Vowells, Hubert, 123-24, 127, 129, 131-32, 238n

Wageworkers' frontier, concept of, 7, 10, 32, 199n
Wagner Act. *See* National Labor Relations Act
Wagoner, Okla., 31
Wallace, Tom, 150
Waller, H. C., 23, 205n
Walton, John C., 116, 150, 166-67
Walton, W. H., 107, 230n, 232n
Wanette, Okla., 16
War Labor Board, 143
Washington, D. C., 77, 90

Washington State, 45, 62, 90, 125
Waynoka, Okla., 148
Wear, Homer, 167-68, 170-73
Weinstein, Della, 19
Welding, 65, 163, 175-76
Wells, Nick, 155
Wermke, Frank, 113-14
Western Association of Free State Employment Bureaus, 46
Western Federation of Miners (WFM), 15-16, 19, 28, 30, 32, 122, 171, 208n
Western Labor Union (WLU), 15
West Virginia, 168
Wewoka Capital-Democrat, 134
Whalen, John, 97, 106, 124, 232n, 237n
Wheat farming, 7-8, 13, 32-33, 35, 37, 45-46, 153
Wheat or Grain Belt, 11, 36-38, 42-43, 49, 52-53, 55-56, 61, 145, 163, 179-82, 246n
Whiting, Ind., 59
Whitt, O. H., 168
Wichita, Kans., 3, 7, 52, 130, 185, 246n
Wichita Falls, Tex., 60
Wier, Henry W., 190
Wiggins, John E., 84, 90, 226n
Williams, Robert L. (Okla. governor), 97
Willis, John G., 27
Wilson, Ollie S., 22
Wilson, William B., 22
Wilson, Woodrow, 77, 88, 90, 96, 101, 113, 136, 143
Wilson and Company, 156
Winnipeg, Canada, general strike, 127
Wobblies. *See* Industrial Workers of the World
Women, 19, 32, 36
Wood, Robert and Ina, 193
Woodward, Okla., 47-48, 132
Woofter, John, 99, 101

Workers, 8, 12, 81, 92; dequalification or deskilling of workers, 5; semiskilled or unskilled, 5, 8; skilled, 5; urban, 180
Workers Defense Union, 123
Workers International Industrial Union (WIIU), 29, 30
Working Class Union (WCU), 31, 42, 77–80, 84–92, 99, 113, 140–41, 155–56, 224–25n
Workingmen's Union of the World (WUW), 188, 256n

Work Peoples College, Duluth, Minn. (IWW school), 150
World War One, or First World War, 7, 28, 39, 57–58, 87, 93, 158, 163, 185, 193–94, 238n
World War Two, 192
World War Veterans, 156
Wynona, Okla., 145

Yale, Okla., 66

Zeigler, C. C., 22, 205n

www.ingramcontent.com/pod-product-compliance
Lightning Source LLC
Chambersburg PA
CBHW020744160426
43192CB00006B/243